ENVIRONMENTAL STRATEGIES FOR SUSTAINABLE DEVELOPMENT IN URBAN AREAS

Environmental Strategies for Sustainable Development in Urban Areas

Lessons from Africa and Latin America

Edited by
EDESIO FERNANDES
Institute of Commonwealth Studies
University of London

Studies
in
Green Research

LONDON AND NEW YORK

First published 1998 by Ashgate Publishing

Reissued 2018 by Routledge
2 Park Square, Milton Park, Abingdon, Oxon, OX14 4RN
711 Third Avenue, New York, NY I 0017, USA

Routledge is an imprint of the Taylor & Francis Group, an informa business

Notice:
Product or corporate names may be trademarks or registered trademarks, and are used only for identification and explanation without intent to infringe.

Publisher's Note
The publisher has gone to great lengths to ensure the quality of this reprint but points out that some imperfections in the original copies may be apparent.

Disclaimer
The publisher has made every effort to trace copyright holders and welcomes correspondence from those they have been unable to contact.

ISBN 13: 978-1-138-33878-4 (hbk)
ISBN 13: 978-1-138-33879-1 (pbk)
ISBN 13: 978-0-429-44144-8 (ebk)

Contents

v

List of contributors

Christian Arandel is Program Advisor at Environmental Quality International-EQI, a leading private consulting firm in Egypt. He manages the Urban Management Program, a joint UNDP-UNCHS-World Bank initiative, in the Arab States Region.

Carl R. Bartone is Principal Environmental Engineer at the Urban Development Division of the World Bank, based in Washington, DC. Prior to joining the Bank in 1986, he spent 15 years working in Latin America as regional pollution control advisor and technical director of the Pan-American Center for Environmental Engineering-CEPIS in Lima, Peru.

Jochen Eigen is the Coordinator of the Sustainable Cities Program of the UN Center for Human Settlements and the UN Environment Program, based in Nairobi, Kenya. He also coordinates the Environmental Component of the Urban Management Program.

Edesio Fernandes is Senior Research Fellow at the Institute of Commonwealth Studies of the University of London. His research interests include urban and environmental planning, policy and legislation in developing countries.

Herbert Girardet, a UN Global 500 Award recipient, is a film-maker and writer; he is the author of *The Gaia Atlas of Cities* and co-author of *Making Cities Work*. He is Visiting Professor for Environmental Planning at Middlesex University, London, and Chairman of the Schumacher Society in the UK.

Leonardo Martínez-Flores works at the Instituto Nacional de Ecología in Mexico City, and is directly involved with the formulation and implementation of the Air Pollution Program.

Janice Perlman, a Kennedy Professor of the University of California at Berkeley, is the founder and Board President of the Mega-Cities Project, which is a transnational non-profit network created in New York in 1987, dedicated to transferring innovative solutions to common problems between the world's largest cities.

Jonas Rabinovitch is the Manager of the Urban Development Team of the United Nations Development Program, based in New York. He worked for the Curitiba Research and Urban Planning Institute for many years.

Bolivar Ruiz-Adaros is the Regional Director of the National Commission for the Environment-CONAMA in the Bio Bio Region, Chile; he is also a lawyer and Lecturer in Environmental Law at the University of Concepcion, in Chile.

David Satterthwaite is the Director of the Human Settlements Program at the International Institute for Environment and Development, based in London, and Editor of the Journal *Environment and Urbanization*.

Vuyiswa Tindleni is a member of the Green Coalition, from Cape Town, South Africa. The Green Coalition is an alliance of over forty NGOs which have been working closely together with the Cape Town City Council on several issues regarding sustainable development.

Acknowledgments

.

This book contains a selection of the papers presented at the International Conference "Environmental Strategies for Sustainable Development in Urban Areas: Lessons from Africa and Latin America", which was jointly promoted in September 1996 by the Institute of Commonwealth Studies-ICS and the Institute of Latin American Studies-ILAS of the University of London. Special thanks are due to all those who accepted the invitation to present papers at the Conference and thus to share their valuable experience with the other speakers and participants: Christian Arandel, Martha Banuelos, Carl Bartone, Jochen Eigen, Herbert Girardet, Mohamed Halfani, Patricia Iturregui, Davindar Lamba, Leonardo Martínez-Flores, Alexandrina Sobreira de Moura, Janice Perlman, Jonas Rabinovitch, Bolivar Ruiz-Adaros, David Satterthwaite and Vuyiswa Tindleni.

At the ICS, I would like to thank Professor James Manor, Ms. Sonja Jansen and Ms. Rowena Kochanowska for their support and help. At the ILAS, I would like to thank Professor Victor Bulmer-Thomas for his encouragement.

I am especially indebted to the Citicorp Foundation and to numerous Citibank branches in Africa and Latin America for their generous grants, which made the Conference possible; I would especially like to thank Mr. Peter Thorpe, Mr. Robert Annibale and Ms. Kathryn Carassalini. Finally, I am also very grateful to Ms. Mira Knezevic and Mr. Pedro Alves for their help with the preparation of the manuscript.

1 Learning from the South

Edesio Fernandes

Introduction

The rapid urbanization in African and Latin American countries throughout this century has provoked all sorts of drastic social, economic and cultural changes, which have already been discussed extensively. Equally well known are the long standing technical, financial and political problems and constraints confronting the governments of those countries in their attempts at promoting economic development and controlling urban growth.

Special emphasis has been placed, since the 1980s, on the complex nature and implications of the processes of (re)democratization and decentralization in Africa and Latin America. The research on the reform of the state in those regions has been combined in recent years with a discussion on the conditions and consequences, for the national economies, of structural adjustment, liberalization and privatization policies imposed by international donors and multilateral financial institutions, within the context of the current process of economic globalization.

However, although the pace and the pattern of overall development in Africa and Latin America has also brought about major environmental changes, the existing research on the environmental realities in those regions is still incipient, particularly as regards the urban environment. In fact, most of the interest in, and knowledge on, environmental matters seems to focus primarily on issues concerning wildlife or the great eco-systems such as the Amazon, while the understanding of the environmental dimension, and resulting problems, of the process of intensive urban growth is still fragmented. Moreover, as Satterthwaite argues in Chapter 4, the research agenda on environmental matters in

1

cities in the South has been determined, to a large extent, by the values, concerns and interests of more developed countries in the North.

Furthermore, it is also fair to say that much of the existing research has had a negative approach, be it when it emphasizes how economic development in the South has reproduced the traditional predatory patterns of the North, or be it when it stresses how the current stage of international division of labor is deeply based upon the maintenance of criteria of environmental injustice concerning the conditions of exploitation of natural resources, as well as the overall distribution of the costs and benefits of development.

Nevertheless, the growing concern about the deterioration of the urban environment in African and Latin American countries has been translated in recent years into several original, promising experiences of urban environmental management which are, as yet, not widely known. This book seeks to present, discuss and evaluate some such experiences. It has two principal aims: to acknowledge the significant advances in environmental management which are being achieved in many African and Latin American cities, and to provide information on some of the main experiences in progress, especially the fundamental ideas behind their formulation, so that they can be replicated in other cases.

Although this book is an edited collection of articles by different authors, it should be viewed as the product of a dialogue of several orders, namely: a dialogue between academics and public administrators; between governmental environmental agencies and environmental Non-Governmental Organizations-NGOs; and between representatives from local, regional, national, transnational and international organizations and institutions.

Above all, besides proposing a dialogue between countries from the South and their richer counterparts in the North, this book also promotes a long overdue dialogue between South and South, particularly between Africa and Latin America. These are two regions which, sharing as they do so many similarities, can only benefit from direct contact with one another - without the often distorting mediation and articulation provided by international organizations with their own agendas. By bringing the experiences of urban environmental management in Africa and Latin America to the attention of a wider audience, this book also aims to involve new participants in this debate.

This book comprises a broad-ranging collection of empirical studies, some of which are more general, others more specific. They touch on issues as diverse as air pollution in Mexico City and garbage collection in Cairo, tree planting in Rio de Janeiro and urban gardening in Accra, public transport in Curitiba and the pollution of the bays of Talcahuano, water supply in Cape Town and the protection of historical heritage in Tunis. However, there is at least one major theme linking the different chapters: that of the need for the formulation of new strategies for urban environmental management which reconcile economic development with environmental preservation in a sustainable manner.

This is not a guide to "best practices", as, with all their undeniable merits, the experiences discussed in this book have many inherent problems and shortcomings. However, while the various problems concerning the replicability of unique experiences are discussed by Rabinovitch in Chapter 2 and Perlman in Chapter 7, among others, the general argument of this book is that some fundamental ideas and concepts are basically right - and ideas are, or can be, replicable.

Sustainable environmental management in the South: an overview

The phenomenon of intensive urban growth in Africa and especially in Latin America has been extensively analyzed since the 1970s, and urban research has progressed enormously in its efforts to understand the factors, agents and processes shaping urban reality in those regions. It is particularly the case of the line of research which proposes a critique of the traditional (neo)liberal framework however dominant that may still be. Within the field of critical urban research, the urban phenomenon has been considered as the dynamic result of a complex and contradictory process of articulation of economic, political, legal and cultural forces, through which both cities and rural areas in Africa and Latin America have been redefined the by the changing nature of capitalism throughout this century, and more recently by the process of economic globalization.[1] Given the complex nature and implications of the processes of (re)democratization and decentralization in those regions, special emphasis has been placed since the late 1970s on the role played by a variety of social and political agents and institutions, especially the state and urban social movements and NGOs.[2]

Parallel to the progress of urban research, a whole tradition of environmental studies has been established since the 1972 United Nations Conference in Stockholm, but only in recent years have such studies focused on urban areas, particularly those in Africa and Latin America.[3] Amongst the main changes underlying the evolution of urban environmental research, the adoption of a wide concept of the urban environment should be stressed: it has increasingly been viewed as the balance between natural, artificial and cultural elements which makes for a determined quality of urban life. The emphasis has been placed on

3

environmental processes rather than on their specific aspects. While focusing on urban areas, environmental studies have also taken into account the environmental dimension of the socio-economic processes taking place in rural areas, particularly given the intertwined relationship between cities and the countryside.[4]

The multifaceted nature of environmental processes has been reflected in the undertaking of original interdisciplinary environmental studies. The growing number of studies and sources of data - UNEP, UNCHS, World Bank, international NGOs (Friends of the Earth, Greenpeace, etc.) and other national and international organizations - should also be mentioned, as they have certainly contributed to increasing the level of awareness of environmental realities in Africa and Latin America.

In particular, as discussed by Eigen in Chapter 9, the role of the UN agencies in raising environmental matters, producing data and information, and stimulating the adoption of original management experiences has been fundamental. Following the 1992 United Nations Conference on Environment and Development in Rio de Janeiro, in which the concept of Agenda 21 was proposed, an intensive mobilization process involving governmental agencies and urban and environmental social movements and NGOs led to the long expected meeting of the "green" and the "brown" agendas during the 1996 "City Summit", the Habitat II Conference promoted by the United Nations in Istanbul. The implementation of the principles of Agenda Habitat is currently the main challenge for cities in Africa and Latin America.[5]

With the development of environmental research over the last decade, many significant changes have taken place, from thematic research to the identification of global phenomena (*El Nino*, global warming, etc.), and from a naturalist to a human-centered approach to environmental processes.[6] Once again, many such studies have reinforced the need to integrate the "green" and the "brown" agendas. In that respect, it is important to remark that most urbanization studies have been undertaken in the context of industrialization resulting from the traditional international division of labor, and therefore the pollution forms discussed have been those traditionally associated with industrialization processes. The factors considered by such research tradition have included, among others, the growing scale of production, population growth, the emergence of social needs, and the use of inadequate technology.

More recently, the new international division of labor supporting post-industrial capitalism has also reinforced urbanization as the condition for the economy of services and information, but this time applying new technologies and producing new forms of pollution. Such different stages of capitalist production can sometimes co-exist in the same country, as has happened in Africa and especially in Latin America.[7]

Another important change concerns the false question of "preservation *vs.* development", and the evolution of concepts in that respect has been clearly reflected in policy and legislation. Following an initial stage characterized by the lack of information and concern, in its second stage environmental research opposed preservation to economic development; more recently, a third stage has been established, in which preservation has been viewed as the very condition for sustainable development. The real challenge for developing countries, therefore, is to view whatever form of pollution as a sign of the inefficiency of the economic model as a whole.[8]

Underlining this conceptual evolution is another most significant change of approach: that from a moral-humanitarian to a socio-political approach to environmental processes in Africa and Latin America, and, more recently, to a socio-economic approach. The truth is that all such approaches are valid in their own right, but need to be combined to be effective and just.[9] The increasing incorporation, by many national laws, of the requirement for an assessment of environmental impact prior to the official authorization of several urban activities reflects such an evolution: as Martinez-Flores argues in Chapter 6, there is a growing awareness that we need to better understand the complex distribution of costs and benefits resulting from economic activities, which can only be achieved through a deep analysis proposing to evaluate the socio-political and cultural as well as the economic costs of pollution.

As was mentioned above, an important tradition of environmental studies has associated the process of globalization of the economy and the conditions of environmental injustice brought about by the new economic and political international relations, which have had a major impact on developing countries.[10] As argued by Satterthwaite in Chapter 4, while the North-South divide becomes even more evident when the environmentally-unfriendly lifestyles in developed countries are considered, the alarming urban health indicators in developing countries prove that the link between poverty and the environment cannot be ignored any longer.[11] Recent events in South-East Asia and in the Brazilian Amazon, amongst many other examples, have also shown the negative environmental impact of the movement towards "economic development at any cost" in developing countries through the pressure to rapidly develop industry, occupy new areas, and burn forests.

While the processes of democratization and decentralization in Africa and Latin America are far from being consolidated, the widespread adoption of liberalization policies has forged new domestic roles for the

5

state in those regions, new relations between states, markets and societies, and new relations between national states. One of the results of such changes has been the gradual definition of a new scope for urban and territorial planning, the focus of which has been shifted from comprehensible to strategic planning, in which the joint action of the public and the private sectors is stimulated, as has happened in the experiences discussed by Bartone in Chapter 5.[12] At the same time, as discussed by Girardet in Chapter 11, the provision of traditional public services has been challenged by newly adopted privatization policies, and the problems resulting from the lack of regulation, combined with the new forms of exploitation of natural resources, have also provoked significant environmental impact.

It is in this changing, and contradictory, context that the discussion of new strategies for environmental management has become essential to the promotion of socio-economic and urban development in Africa and Latin America.

The long standing problems of urban and environmental management in those regions are widely known; among them, the lack of resources and coordinated action between governmental agencies has frequently been mentioned. Many analyses have criticized the fragmented view which has characterized the environmental action of such agencies, their lack of a clear policy and specific objectives, as well as the confused, if not conflicting, distribution of legal and politico-institutional powers prevailing in many countries.[13] If the growing literature on development and urban studies in Africa and Latin America has also increasingly incorporated an environmental dimension, and the meeting of the "green" and "brown" agendas seems indeed to be taking place at last, as Ruiz-Adaros argues in Chapter 8 in many countries the legal-institutional apparatus in force still needs to be reformed in order to integrate both agendas.

Combining the conclusions from development and urban studies, the research on environmental management in African and Latin American countries has stressed the importance of the incorporation of new principles into traditional management strategies, including the adoption of inter-sectoral approaches, the notion of partnership, and the emphasis on the need for participation, transparency and accountability to guarantee the legitimacy and efficiency of the decision-making process. The need for a broad approach to the complex urban environmental realities is of utmost importance, and it depends on a proper integration between policy-making, institutional organization and effective management.

As Arandel argues in Chapter 10, given the complexity of the problems involved, the formulation of environmental strategies for sustainable development in urban areas can only lead to effective actions if, expressing a concern for the overall environment, they rely on the proper definition of objectives and on the adequate distribution of responsibilities among the multiple agents, public and private, involved in the process.

Several manifestations of this new approach to the management of the urban environment have been formulated, and experimented, in many African and Latin American cities. As the examples discussed in this book clearly indicate, some are more consolidated and successful than others, some are promising but as yet still unsatisfactory, while others have a long way to go before they can make a real difference. But, it is only fair to acknowledge the fact that there is a movement of change happening in those regions, and that there is indeed cause for hope for better future prospects than those expressed in so many bleak analyses of the "dark continents". There is more to Latin America, for example, than the internationally acclaimed experience of Curitiba, however significant it is. In fact, there is more to Curitiba itself than what is widely known, as Rabinovitch discusses in Chapter 2.

It could be argued that, even given due respect to the enormous extent of their existing financial, social, urban and environmental problems, African and Latin American countries are in a privileged position: besides being able to benefit from the experience and technical know-how already available in developed countries, and therefore being in a position to minimize their past mistakes, perhaps they may have the unique possibility of approaching these problems differently. This is due to the fact that, as fundamental political institutions are still being created in those regions on a daily basis, people can be more directly agents of their own history than those living in richer countries, where consolidated political institutions and mechanisms offer less room for change. That is the reason why the most successful experiences presented in this book stress the fundamental importance of all forms of participation, especially the political participation in the decision-making process, for the effective formulation and implementation of environmental strategies for sustainable development in urban areas.

In other words, political history is in the making in the South, suggesting the possibility for more and new opportunities. Above all, what the experiences discussed in this book reveal is that the truly unique, still unexplored resource in the countries in the South is their people, whose history of strength-as-resiliency must be transformed into a more active and creative civic force. As Tindleni indicates in Chapter 3, perhaps no other country symbolizes this notion more than South Africa.

To put it briefly, African and Latin American countries have a great deal to learn from the North as to what alternative actions can be followed and, above all, what mistakes should not be repeated. They can only benefit from talking directly to, and learning from, each other: there are many examples that reveal that much can be done with limited means. But, as the experiences discussed in this book demonstrate, perhaps the North could also learn a thing or two from the South.

The structure of the book

The structure of this book reflects is main objective: to promote a dialogue between different approaches to the matter of sustainable environmental management in urban areas. Given our belief in the need for an interdisciplinary, multifaceted and cross-sectoral understanding of urban environmental development and management, the contributors include environmental researchers and activists, academics, government officials, members of local, transnational and international NGOs, as well as members of international development organizations and multilateral financial institutions. They include leading professionals and international specialists in the field, and many of their earlier publications have established the basis for our understanding of the subject matter of this book.

The book is international in scope. While significant cases from nine countries are examined in some detail - Brazil, Chile, Mexico, Egypt, Ghana, South Africa and Tunisia - some authors also draw on their professional experience and previous research in a wide range of other developing and developed countries. This allows them to go beyond the specificities of the case studies to identify the main characteristics and implications of environmental management in urban areas. It also enables them to identify the main trends in policy, and the main themes for future research and practice.

It is important to stress that, as it seems to be increasingly happening in the wider context of environmental research and practice, despite their different backgrounds the contributors to this book do speak the same language; the different formal treatments they give to their experience and ideas have been kept to some extent.

In Chapter 2, following a discussion of the general principles guiding the activities of the United Nations Development Program aiming at promoting urban sustainability, within the context of the current climate favoring local-global linkages, Jonas Rabinovitch describes the very successful experience of urban environmental management of the city of Curitiba, in Brazil. Principles and procedures adopted in the city are examined in detail, especially from the viewpoint of the many lessons other cities could learn from Curitiba. Rabinovitch is aware of the problems of replicability, but he firmly believes that, if common sense and serious, continuous work prevail, many of the principles and strategies responsible for the city's successful experience of urban environmental management can be repeated in other cities. Moreover, they can also be extended to rural areas, as has happened in the State of Parana (of which Curitiba is the capital city), in order to support an original policy of integrated "rurban" development

In Chapter 3, Vuyswa Tindleni discusses the role of NGOs and community organizations in the formulation of an original, integrated and democratic sustainable development strategy for Cape Town, South Africa, within the wider context of the national Reconstruction Development Program. She stresses the importance of the partnership between government and civil society at all stages, from the identification of problems to the development of an action plan, but warns that the effective formulation and implementation of plans depends on the extension and strengthening of participatory democracy, especially at the local level.

Chapter 4, by David Satterthwaite, discusses what he considers to be the main environmental problems in cities in the South, which have been largely neglected by researchers as they do not fit the agenda determined by Northern perceptions - which favors issues related to resource use and waste or ambient air pollution, to the detriment of the many categories of problems that are the main causes or contributors to premature death, illness and injury in the South. Following a discussion on how a combination of social, economic and political factors determine who is most affected by such problems, as well as who benefits from them, Satterthwaite then discusses the links between environmental problems and the size, density and rate of growth of cities, also relating them to the existing levels of income in the different urban centers. He strongly emphasizes the need to draw on the knowledge and resource of low-income communities, which can only happen if the broader political process actually devolve powers - and not only responsibilities - to the community level.

In Chapter 5, Carl Bartone discusses the main principles behind the World Bank's participation in the formulation of urban environmental strategies, namely: extending basic urban environmental services, especially for the urban poor; reducing pollution that imposes high costs

9

on cities; and building sustainable institutions for managing the urban environment. Emphasizing the need to enable participation, build commitment and strengthen local capacity, he presents two cases in which such principles have been successfully applied, the formulation of a watershed protection plan in São Paulo, in Brazil, and the strategic sanitation plan for Kumasi, in Ghana.

The new program for tackling the alarming, and highly complex, air pollution situation in Mexico City is presented in Chapter 6, by Leonardo Martinez-Flores. He stresses the point that only a new conceptual approach, systemic and integrated, relating environmental quality to several urban processes, can lead to the formulation of appropriate criteria to identify strategies and policy instruments; above all, he argues that the achievement of fundamental solutions depends on profound cultural changes in the relationship we currently maintain with the city and the overall environment.

In Chapter 7, Janice Perlman discusses three cases in which, as identified by the transnational Mega-Cities Project, effective and innovative solutions have been given to local environmental and poverty problems, namely, the "Zabbaleen" garbage recycling and micro-enterprise development in Cairo, Egypt; the Paid Self-Help Reforestation Project in Rio de Janeiro, Brazil; and the Urban Market Gardens project in Accra, Ghana. Again, despite the abovementioned problems of replicability, Perlman argues that many cross-cutting lessons can be drawn from such unique experiences, especially concerning their implementation, which fundamentally depends on multi-sectoral partnership, direct community participation, and the formulation of integrated strategies.

In Chapter 8, Bolivar Ruiz-Adaros discusses the experience of environmental management in the region of Bio Bio, in Chile, particularly the principles behind the Environmental Recovery Plan for Talcahuano, which is a fine example of both regional management and multi-sectoral partnership. He places emphasis on the importance of an adequate legal-institutional framework for the effective formulation and enforcement of environmental strategies, stressing that they contribute to, and depend on, profound cultural changes.

In Chapter 9, Jochen Eigen presents the fundamentals of the Sustainable Cities Program, which is a joint UNEP/UNCHS program that has already been applied in many cities in Africa and Latin America. Being based upon the key principle of broad-based local governance, this UN Program proposes to improve the conditions of urban environmental planning and management capacity. Although driven by local needs and opportunities, the Program provides a framework for linking local actions at the national, regional and global levels. Inter-agency co-operation and broad partnership schemes are vital to the success of these experiences.

Chapter 10, by Christian Arandel, discusses innovative strategies for sustainable development in the Middle-East and North Africa in three key areas identified by Environmental Quality International, namely the expansion of small businesses, the promotion of informal sector participation in the environmental services industry, and the revitalization of historic city centers. Successful examples concerning the Alexandria Business Association, in Egypt, the abovementioned Zabbaleen program, in Cairo, and the revitalizing of the Medina of Tunis, in Tunisia, illustrate each of those principles, which draw from, and aim to reinforce, the region's entrepreneurial spirit and rich history. Arandel also discusses the importance of incorporating issues concerning gender and the media into the consideration of the abovementioned principles, so that urban environmental policy can effectively address contemporary needs and future challenges.

Finally, in Chapter 11, Herbert Girardet proposes a general analysis of the implications the huge resource use by modern cities has for both local and global environments. Comparing cities to superorganisms, he describes their intense metabolism and assesses their ecological footprint; the example of London is significant and lessons should be learned from it, not least because the city pioneered large-scale urban development and became an international center for financial services, but also because its footprint reaches unique proportions. Girardet then suggests what the conditions for sustainable development should be, and proposes policies for the United Kingdom which, if implemented, could have an impact on cities elsewhere, as well as reducing their footprint. Above all, he believes in the need for the adoption of a new paradigm for a renewed urban culture in which sustainability, peace and personal empowerment are reconciled. However, he argues that the success of cities requires the extension of popular participation in strategic decision-making to strengthen local democratic processes.

Suggested themes for further research

As I have stressed, this book aims to widen the debate on urban environmental management in Africa, Latin America and elsewhere. While the contributors have addressed several of its main aspects, I hope that the studies presented here stimulate further research into these and other issues.

Among the themes which deserve better understanding, I believe that more empirical studies should focus on the importance of the legal system - both formal and informal - for the adequate formulation and effective implementation of environmental strategies for sustainable development. The role of law in this process must be questioned, including the phenomenon of widespread illegality among the urban poor - and even among the rich - especially as to their conditions of access to land and housing, which certainly leave a major imprint on the environment.[14] In fact, the integration of urban and environmental laws has not been always straightforward, partly due to the fragmented institutional apparatus which, as many of the studies in this book reveal, prevails in many countries.

The environmental implications of state intervention, through law, in the process of urban growth and change, require further research. The potential conflict between land legalization policies and environmental protection, for example, deserves to be better understood, especially as the environmentalist stance has been frequently espoused by conservative interests opposed to regularization of land tenure - as has happened in Brazil.[15]

The issue of property rights is central to much of this debate, but it begs another question concerning an apparent paradox. Many of the strategies for urban environmental management discussed here underline the need to redefine traditional urban property rights and ideologies in order to make room for a new approach, widening the scope for urban policy to improve living conditions. However, most developing countries have recently been undergoing significant political and administrative reform in the context of economic liberalization and privatization policies. How can such policies be reconciled with a more progressive, environmentally-friendly, notion of property rights?

Last, but not least, I would stress the crucial importance of a deeper reflection on the issue of governance. One the one hand, as Ruiz-Adaros suggests in Chapter 8, emphasis should be placed on examining the relations between central and local governments in African and Latin American countries, particularly the implications of decentralization for urban environmental management.[16] On the other hand, and more importantly, we need to achieve a better understanding of recent trends in urban governance, including the experiences of popular participation in the decision-making and budgeting processes, and their implications for the reform of the urban and legal-institutional orders.[17]

The papers presented here demonstrate that, given the nature of the relationship between politics, urbanization and the environment, any significant change in the approach to urban environmental management is only to be attained through a more encompassing and democratic political process.

Notes

1 See, among others, Gilbert (1996), Smith, D. (1992), Becker et al (1994) and Wekwete & Rambanapasi (1994).
2 See, for example, Edwards (1995) and Bradford, Jr. (1994).
3 See, among others, Collinson (1996).
4 See, among others, Hall (1997); I have also discussed some aspects of the complex question of the preservation of the Brazilian Amazon elsewhere (Fernandes, 1996a).
5 Among the many valuable publications by UN agencies, see UNCHS (1996), UNCHS (1997) and UNCHS/UNEP (1997).
6 See, for example, the studies collected by Serageldin et al (1995).
7 See Castells (1989).
8 See, among others, Badshah (1996) and Burgess et al (1997).
9 See Goldblatt (1996), Hurrell & Kingsbury (1992) and Collinson (1996), for a discussion of environmental politics in Brazil, see Guimaraes (1991).
10 See Borja & Castells (1997).
11 See, among others, Badshah (1996) and Burgess et al (1997).
12 See Harris & Fabricius (1996) and Healey et al (1997).
13 I have discussed aspects of the environmental management in Brazil elsewhere (Fernandes, 1996b).
14 For a collection of critical studies on the relation between law and urbanization in developing countries, including Brazil, Mexico, Venezuela, South Africa and Kenya, among others, see Fernandes & Varley (1998); I have discussed the Brazilian case in detail elsewhere (Fernandes, 1995a).
15 I have discussed this problem within the wider scope of a reflection on the conditions of access to judicial power in Brazil; see Fernandes (1994; 1995b).
16 See, for example, Gilbert et al (1996) and Devas & Radoki (1993); see also Souza (1997) for a detailed account of the political-financial dynamics of decentralization in Brazil.
17 Many such experiences are taking place in Brazil; see, for example Genro et al (1997) and Figueiredo & Lamounier (1996); I have discussed the original experience of Belo Horizonte's participatory budget elsewhere (Fernandes, 1996c).

References

Badshah, A. (1996) *Our Urban Future: New Paradigms for Equity and Sustainability*, Zed Books: London.

Becker, C.M. et al (1994) *Beyond Urban Bias in Africa: Urbanization in an Era of Structural Adjustment*, Heinemann: Portsmouth.

Borja, J. & Castells, M. (1997) *Local & Global - The Management of Cities in the Information Age*, Earthscan: London.

Bradford Jr., C.I. (1994) *Redefining the State in Latin America*, OECD: Paris.

Burgess, R. et al (eds) (1997) *The Challenge of Sustainable Cities: Neoliberalism and Urban Strategies in Developing Countries*, Zed Books: London.

Castells, M. (1989) *The Informational City: Information Technology, Economic Restructuring and the Urban Regional Process*, Blackwell: Oxford.

Collinson. H. (ed) (1996) *Green Guerrillas: Environmental Conflicts and Initiatives in Latin America and the Caribbean*, Latin America Bureau: London.

Devas, N. & Radoki, C. (eds) (1993) *Managing Fast Growing Cities*, Longman: London.

Edwards, S. (1995) *Crisis and Reform in Latin America*, Oxford University Press: Oxford.

Fernandes, E. (1994) "Defending collective interests in Brazilian environmental law: an assessment of the Civic Public Action", in 3 *RECIEL - Review of European Community and International Environmental Law*.

Fernandes, E. (1995a) *Law and Urban Change in Brazil*, Avebury: Aldershot.

Fernandes, E. (1995b) "Collective interests in Brazilian environmental law", in Robinson, D. & Dunkley, J. (eds) *Public Interest Perspectives in Environmental Law*, Wiley Chancery Law: Chichester.

Fernandes, E. (1996a) "Environmental zoning: a solution for the Amazon Region?", in 5 *RECIEL - Review of European Community and International Environmental Law*.

Fernandes, E. (1996b) "Constitutional environmental rights in Brazil", in Anderson, M.R. & Boyle, A.E. (eds) *Human Rights Approaches to Environmental Protection*, Oxford University Press: Oxford.

Fernandes, E. (1996c) "Participatory budget: a new experience of democratic administration in Belo Horizonte, Brazil", in 7 *Report*.

Fernandes, E. & Varley, A. (1998) *Illegal Cities - Law and Urban Change in Developing Countries*, Zed Books: London.

Figueiredo, R. & Lamounier, B. (1996) *As Cidades que dao certo: experiencias inovadoras na administracao publica brasileira*, MH Comunicacao: Brasília.

Genro, T. et al (1997) *Desafios do Governo Local: o modo petista de governar*, Fundacao Perseu Abramo: São Paulo.

Gilbert, A. (ed) *The Mega-City in Latin America*, United Nations University Press: Tokyo, New York, Paris.

Gilbert, R. et al (1996) *Making Cities Work*, Earthscan: London.

Goldblatt, D. (1996) *Social Theory and the Environment*, Polity Press: Oxford.

Guimaraes, R.P. (1991) *The Ecopolitics of Development in the Third World: Politics and the Environment in Brazil*, Lynne Rienner Publishers: Boulder and London.

Hall, A. (1997) *Sustaining Amazonia: grassroots action for productive conservation*, Manchester University Press: Manchester and New York.

Harris, N. & Fabricius, I. (1996) *Cities & Structural Adjustment*, University College London: London.

Healey, P. et al (1997) *Making Strategic Spatial Plans*, University College Press: London.

Hurrell, A. & Kingsbury, B. (eds) *The International Politics of the Environment*, Oxford University Press: Oxford.

Serageldin, I. et al (1995) *The Human Face of the Environment*, The World Bank: Washington, D.C.

Smith, D. (ed) (1992) *The Apartheid City and Beyond*, Routledge: London.

Souza, C. (1997) *Constitutional Engineering in Brazil: the politics of federalism and decentralization*, Macmillan: Houndmills and London.

UNCHS (1996) *An Urbanizing World: Global Report on Human Settlements*, Oxford University Press: Oxford.

UNCHS (1997) *The Istanbul Declaration and Habitat Agenda*, UNCHS: Nairobi.

UNCHS/UNEP (1997) *Implementing the Urban Environment Agenda*, UNCHS/UNEP: Nairobi.

Wekwete, K.H. & Rambanapasi, C.O. (eds) (1994) *Planning Urban Economies in Southern and Eastern Africa*, Aldershot: Avebury.

2 Global, regional and local perspectives towards sustainable urban and rural development

Jonas Rabinovitch

Introduction

It is well known that the end of this century will witness an unprecedented challenge in the development of human settlements: for the first time in history, more people will live in cities and towns than in rural areas.[1] During the past three decades, the urban population of developing countries has tripled. By the year 2000, some 2.2 billion people will live in the urban areas of Asia, Africa and Latin America, with approximately half of them in cities of a million inhabitants or more. At least fifty cities will be homes to more than four million inhabitants each.[2] More than simply a demographic phenomenon, rapid urbanization is one of the most significant processes affecting developing countries and shaping their future. The immediate result is a radical transformation in the structure of cities, accompanied by complex social, economic and environmental changes.

This paper aims to introduce, in general terms, some elements of the changing international scenario which is becoming more favorable to enhance local-global linkages, with emphasis placed on urban development issues. It describes current international urban-related initiatives, especially the urban development cooperation strategy of the United Nations Development Program-UNDP in view of such international trends, and it outlines some key elements conducive to sustainable urban development practices.

I shall then briefly analyze the experience of the city of Curitiba, in Brazil, as a basic reference; the incipient transfer of Curitiba's experience to the whole State of Parana will also be briefly described, with emphasis on rural-urban linkages, followed by a discussion of some issues related to the replicability of successful urban management experiences. Put briefly, the paper attempts to highlight some relevant linkages that, following the conclusions of the 1996 Habitat II Conference, in Istanbul, can reinforce and complement lessons from local, regional and international perspectives.

Local-global linkages: towards a cooperation framework

The international development community is gradually acknowledging the inter-connections between global and local events and initiatives. At the 1992 "Earth Summit" (the United Nations Conference on Environment and Development-UNCED) in Rio de Janeiro, Agenda 21 called for support from local actors, both local and central government and Non-Governmental Organizations - NGOs. Following UNCED, there has been a shift in the perception of human, social and urban development In Copenhagen, at the Social Summit, support for civil society was stressed in order to address the complex issues of social integration, poverty eradication and employment. At the Women's Summit in Beijing, the role of NGOs in promoting gender equity was emphasized. Also in Istanbul, at the "City Summit", was the role of local actors seen as central to building sustainable human settlements.

Overall, humankind no longer seems to accept development at the expense of the poor. In keeping with this vision, UNDP has adopted the concept of "Sustainable Human Development" (SHD) as the basic framework for all its development cooperation. SHD is a form of development that not only generates growth, but distributes its results equitably; it regenerates the environment rather than destroying it, and empowers people rather than marginalizing them. It is development that is pro-nature, pro-poor, pro-women and pro-people.[3]

Apart from demographic aspects, there are many reasons why human settlements are key factors in helping to promote SHD:

1 Cities and towns are seriously affected by overcrowding, environmental degradation, under-employment, social disruption and inadequate housing infrastructure and services.

17

2 The origin of many global environmental problems related to the patterns of production and consumption, waste, air and water pollution is to be found at cities and at the local level.
3 Cities are engines of economic growth, and thus the economic prosperity of nations will also depend on their performance.[4]

Global trends

It can be argued that the international community does recognize the global nature of the above issues, and proposes that a global approach could also greatly enhance progress towards achieving sustainable human settlements.[5] At the same time, the global context presents a series of trends which seem conducive to closer local-global interaction. These are mentioned below for the sole purpose of establishing the broader context, as this paper does not intend to analyze in detail these important trends nor their mutually enforcing inter-linkages.

Decentralization Democratic governments and market-based economic systems have gradually replaced authoritarian regimes in many countries during the past ten years. Out of 75 developing and transitional countries with populations greater than five million, all but 12 claim to have embarked on some form of transfer of political power to local units of government.[6]

The end of a bipolarized world The end of the Soviet Union, the creation of the Commonwealth of Independent States and its consequences are re-designing international relations, financial investment patterns, official development assistance and many other worldwide interchanges. Cities and metropolitan areas emerge as independent units competing for resources and investments, both in developed and developing countries. A group of powerful inter-linked internationalized cities has taken full advantage of the opportunities of the new economic and technological environments. National governments acknowledge the economic and social importance of metropolitan regions, which also emerge as new key players in the international scenario.

Globalization Globalization is another trend that plays an important contextual role in shaping new local-global relations. As a preparatory activity for the Habitat II Conference, UNDP organized the round-table "The Next Millennium: Cities for People in a Globalizing World", which brought together more than 150 development thinkers and experts, who then produced more than 80 papers outlining the challenges and

18

opportunities for human development in urban areas in a globalizing world. Among the various recommendations, the following should be highlighted: existing patterns of urban production and consumption should be made more sustainable; the human development of cities should be improved; social and political governance at the local level should be strengthened; and resources for urban development should be mobilized.

Competition and cooperation between cities If, on the one hand, decentralization and globalization trends seem to place cities and metropolitan areas in new competitive arenas for resources and investments, on the other hand the level of technical cooperation among cities is reaching unprecedented levels. The strengthening of cities' associations and NGOs working directly with local authorities and communities enormously facilitates the process of interchanging information and experience. The problems faced by cities worldwide are alike to some extent, and both policy-makers and urban managers can improve their capacity through mutual learning.

Since the pre-UNCED World Urban Forum, held in Curitiba, Brazil, major global cities' associations have understood the importance of speaking with a unified voice. The International Union of Local Authorities, United Towns, Metropolis, The Summit of the World's Major Cities and a series of other international, regional and national cities' associations have joined forces in the creation of the "Group of Four Plus"[7], a coordinating mechanism which also facilitates contacts between local and international actors. This mechanism was very useful during the preparation of Habitat II and is presently being revised by the cities' associations.

Importance of vertical and horizontal relations Local authorities worldwide have been persistently arguing for stronger vertical linkages between municipal and national powers. However, many municipalities tend to overlook the importance of horizontal relations at the local level: for example, the dialogue between city hall, local councils, NGOs, Community-Based Organizations (CBOs) and other such civil society organizations. At UNDP, this process is referred to as "local-local dialogue", which is supported by programs such as the *Local Initiative Facility for the Urban Environment* (LIFE), presently operational in 14 countries. More details about LIFE will be presented later.

New economic blocks There seems to be a growing tendency for countries to consider, at the same time, the advantages of international economic liberalization - a move towards more open commerce through less tax barriers - and the advantages of protectionism - a move towards the protection of national industries and domestic economies. Both approaches seem to exist simultaneously in the policy agenda of most countries, depending on the issue at stake and depending on their individual import-export profile. The formation of the European Union, NAFTA, MERCOSUL in South America and other economic blocks is a concrete demonstration of this fact. In South America, for example, MERCOSUL has certainly increased economic relations, but it has also began to internationalize urban poverty characteristics. Brazilians are proud to produce 70 per cent of the soya bean crops in Paraguay, despite the fact that several poor families had to cross the border in search of work, ending up homeless and jobless in peri-urban areas after the harvest.[8]

This paper does not intend to analyze these complex relations, but merely to draw attention to the fact that local governments can no longer afford to ignore what happens internationally, no matter how distant these issues may seem to be for the routine of a local administration. Global trends will influence both the patterns of employment and the availability and location of investments at the local level. National welfare-state policies seem threatened, placing an increasing burden on the social agendas of local government.

Technological communications/information revolution The growing flow of information and improved communication facilities affect cities in at least two unprecedented ways. On the one hand, it affects them externally due to factors such as: improved communications with other cities and other levels of government; access to potentially useful technical information and technologies; possibility of stronger linkages between universities, city administrations, private sector and other institutions, among other potential advantages. On the other hand, it also affects cities internally, especially in so far as the following factors are concerned: potential to improve land registry databases and information about the city; improved communications among internal departments; improved public relations; possibility to increase municipal revenues through more detailed monitoring of property taxes, among other potential advantages.

Nowadays, in principle, a student can have as much information as the office of the mayor, which factor can contribute to more democratic urban governance practices. However, it is also important not to mystify technology: software can complement, but it does not substitute human capacity development. All the elements above seem to point to a changing context in which "urban/rural", "global/local", "city/countryside", are not necessarily competing and mutually exclusive aspects, but increasingly complementary in forming a path towards sustainability.

A growing challenge: are cities sustainable?

Environmental consequences of urbanization in developing countries

Urbanization in developing countries has been accompanied by an alarming growth in the incidence of poverty and environmental degradation. Today, one out of four urban dwellers lives in absolute poverty; another one in four is classified as relatively poor. By the end of this decade, according to the UN, poor urban households are projected to increase by 76 per cent. This rapid concentration of hundreds of millions of people has placed an extraordinary strain on the ability of governments - both municipal and national - to meet the needs of city dwellers. Urban environmental problems - such as water and sanitation, waste management, public transport and energy - are particularly acute, and growing worse, as available services and resources are overwhelmed by expanding populations.

Many city dwellers in developing countries live in crushing poverty; by the year 2000, more than half the developing countries' poor will be living in cities and towns: 90 per cent in Latin America, 45 per cent in Asia and 40 per cent in Africa. Their living conditions are alarming, as their numbers far outstrip the supplies of water, waste removal, transport and health assistance.

Urbanization trends are inevitable

The growth of cities is inevitable - and irreversible. Standing at 2.4 billion in 1990, the world's urban population will rise to 3.2 billion in the year 2000 and 5.5 billion in 2025. The developing countries' share in these totals - 63 per cent in 1990 - will rise to 71 per cent in 2000, and 80 per cent in 2025. By the end of the 1990s, Mexico City will have almost 25 million people, São Paulo almost 22 million. Calcutta, Shanghai and Bombay will each have more than 15 million residents, and 13 other cities in developing countries will have more than 10 million, namely: Seoul, Cairo, Dacca, Delhi, Lagos, Beijing, Bangkok, Manila, Jakarta, Karachi, Tianjin, Buenos Aires and Rio de Janeiro.

In addition to the growth of these large scale cities, the growth of small and medium-sized cities will also continue at unprecedented rates.

21

At the same time, the growth of peri-urban areas is reaching considerable proportions, although with different regional characteristics. More recent development plans recognize that cities have multiple and complex links with rural areas and enhance the importance of urban-rural linkages.

Issues of priority concern

In August 1994, the International Colloquium of Mayors on Social Development was held at the United Nations Headquarters in New York. The event was sponsored by UNDP in cooperation with the United Nations Center for Human Settlements-UNCHS and the abovementioned Group of Four Cities' Association. Prior to the Colloquium, a survey undertaken by UNDP with mayors worldwide provided a representative sample of the main problems presently faced by urban areas: unemployment was listed as the most serious urban-related problem, followed by inadequate housing. The survey clearly demonstrated that, despite national differences, urban problems are very similar worldwide.

Others priority issues mentioned by mayors as "most severe" were, in this order: insufficient solid waste management; violence and personal insecurity; urban poverty; inadequate sanitation/sewerage; air pollution; lack of public transport; inadequate water supply; inadequate social services (health, education); insufficient participation; and discrimination (ethnic, women, poor). A sample of 135 mayors geographically distributed throughout North America/Western Europe, Eastern Europe, Latin America and Caribbean, Africa, Arab States, Asia and Pacific answered the questionnaire.

Are cities sustainable?

Perhaps the most appropriate question is: can cities become more sustainable? Intrinsically, cities are not necessarily sustainable. Most local and global environmental problems are generated in cities or are caused by consumption patterns existing in cities. This simple fact is the key to the understanding that cities do have a very important role to play in the path towards sustainability. For example, urban air pollution is a major problem in cities in developing countries, overwhelmingly derived from energy combustion in transport, industry and households in densely inhabited areas. Half of Bangkok's population is reported to suffer from respiratory problems due to the poor quality of the air caused by traffic congestion and inappropriate fuel mixes, while a similar trend is faced by millions of inhabitants in Mexico City.

Energy use is also often the source of wastes that degrade land and water courses. In addition, it is directly related to damage to much wider areas (as with acid rain) and to the atmosphere (greenhouse gases, global warming and ozone depletion). The world's urban wastes (excluding construction debris) are estimated at some 720 billion tons annually, of which 440 billion tons are generated in industrial countries. Only a very small percentage of this amount is dealt with properly, causing drainage problems and breeding ground for infectious diseases, ultimately with serious public health consequences. The Peruvian cholera epidemic that began in Lima and killed 3,600 people in 12 countries in 1991 is a clear example of the dangers of contaminated water supplies, inadequate sewage disposal, irrigation with contaminated water, and poor hygiene in the handling and preparation of food.

At the same time, urban wastes do not only generate local problems. Agenda 21 clearly points out that 70 per cent of maritime pollution is primarily caused by land-based activities, encompassing poor land use management, inadequate sewerage facilities and industrial pollution being brought from inland streams and rivers into international waters.

To think "urban", however, is to think hope - and not despair. Today the growth of cities is increasingly seen as essential for human development. The Gross National Product per capita numbers are much higher in countries with more of their people living in cities. Cities produce 60 per cent of Gross Domestic Product in developing countries. The economics of scale in large cities generates goods and services far in excess of their share of the total population. This higher productivity of urban labor means that wages are higher, while employment opportunities are greater, especially for women. Cities also give their residents the knowledge and skills to become more productive - a beneficial cycle. Cities promote the modernization of agriculture, provide markets for farm goods and reduce pressure on the land.

Moreover, the urban environment permits the efficient mobilization of people's energies and resources, enhances the income-generating ability of the informal sector and promotes education and skills training. Greater participation in the urban economy leads to the creation of a skilled and literate workforce. Thus, the promise of cities is to liberate the mass of people from poverty, hunger, disease and premature death.

What can make cities more sustainable

A deteriorating urban environment is the enemy of sustainable development. Concern for the urban environment is not simply a concern for the health of urban populations or for the hypothetical rights of

23

unborn generations: it is a concern for making sustainable development possible and manageable. A few fundamental points can be mentioned:

1 A more efficient management of the city environment is needed. A concern with reduced use of natural resources per unit of output should be enforced (for example, through public transport policies addressing mobility needs and environmentally sound industrial development), as should policies enhancing savings in public services (for example, a demand-oriented approach to water supply management), increased attention to operation and maintenance of public facilities, improved governance and political will towards sustainability principles.
2 Improved land use planning within urban and peripheral urban areas, thus avoiding the occupation of disaster prone areas and the encouragement of balanced high and low density uses.
3 Adoption of policies that are conducive to a more equitable access of city services (water, energy, transport, waste disposal, educational and health facilities) by the urban poor as well as to more participatory decision-making processes.

What UNDP is doing to promote the sustainability of cities

UNDP has adopted an approach to urban problems that links sustainable human development with economic productivity. It encourages the adoption of strategies that promote equitable growth, gender equality and participatory development. The main goal is to enlarge people's choices by assisting in the development of their capabilities, improving their access to employment, credit, health and education, and increasing their participation in economic, social and political activities. UNDP promotes human development in urban areas by means of providing support for actions in five priority areas, namely: poverty alleviation; strengthening urban local government and administration; improving the urban environment; providing urban infrastructure, shelter and services; and promoting the private sector and NGOs.

As of the first quarter of 1993, UNDP, including the UN Capital Development Fund-UNCDF, was assisting over 280 ongoing urban development and human settlement projects at a total cost of over US$ 470 million. Of this amount, US$ 330 million was provided from UNDP resources and US$ 140 million from government and third party cost sharing contributions. UNDP-assisted programs are executed by specialized agencies of the UN system, and increasingly by developing

24

country governments and NGOs. UNDP's main partner agency in the urban sector is the United Nations Center for Human Settlements - UNCHS/HABITAT, which is promoting the role of cities in sustainable development, strengthening urban management, and coordinating the implementation of the Global Strategy for Shelter in the Year 2000. UNDP has also worked in close collaboration with the United Nations Children's Fund-UNICEF, the International Labor Organization-ILO, the World Health Organization-WHO, the United Nations Population Fund-UNFPA, the United Nations Department of Development Support and Management Services-DSMS, Regional Commissions and the World Bank. Urban infrastructure, shelter and services represent 37 per cent of all ongoing projects.

Some examples of ongoing national, regional and global-level cooperation among partner agencies follow.

Urban Management Program (UMP)

UMP is a ten-year-old global technical support program designed to strengthen the contribution that cities in developing countries make toward human development, including economic growth, social development and the reduction of poverty. UNDP provides the core funding and overall monitoring for the UMP; HABITAT is the executing agency, with the World Bank as an advisory agency. In addition, WHO, ILO, other UN Agencies, bilateral donors and NGOs provide various important types of support. UMP's uniqueness is its capacity to draw on the strengths of the three multilateral partner agencies to create a demand-driven, coordinated approach to technical cooperation in five areas of concentration, namely: municipal finance; land management; urban infrastructure; urban environment; and poverty alleviation.

Local Initiative Facility for Urban Environment (LIFE)

The principal objective of LIFE is to promote "local-local" dialogue - for example, the contact among municipalities, NGOs and CBOs - to improve the quality of the urban environment. Through LIFE, developing countries and multilateral and bilateral donor agencies recognize the crucial role local authorities, NGOs and CBOs play in promoting sustainable urban environment and development. Fourteen countries have been chosen for the pilot phase of the LIFE program. In each selected country a participatory consultation will be held to bring together NGOs, CBOs, local authorities, central government and the private sector, aiming to establish priorities and guidelines for the selection and funding of small-scale projects by a national selection committee. Proposals for regional and interregional activities are submitted by NGO networks and cities' associations.

Public-Private Partnership for the Urban Environment

The purpose of this program is to create a mechanism which can help promote the involvement of the private sector in programs and projects addressing urban environmental problems in developing countries. UNDP has several ongoing urban programs, two of which already enjoy great donor support, the abovementioned LIFE and UMP. Whereas these programs are mainly oriented toward building capacities of governments, municipalities and NGOs, Public-Private Partnerships focus on enhancing the role of the private sector in three specific areas, namely: water and sanitation; waste management; and energy.

A small number of cities will be selected initially. In each of these cities, models of technology cooperation are being established, and it is hoped that these models will be replicated in other cities. The main goal of the projects is the dissemination and exchange of technology/information and the development of concrete joint-ventures between public and private sector agents to tackle the abovementioned problems. UNDP's Project Development Facility for Public-Private Partnerships was successfully launched during Habitat II, initially with the support from the governments of Switzerland, Netherlands and New Zealand.

Ultimately, a successful project is the one that supports the improvement of living conditions of the urban and rural poor in a sustainable manner. Sustainability implies a sensible balance between economic, environmental and social policy interventions. In this context, the experience of Curitiba, Brazil, deserves to be briefly presented to illustrate how these concerns have been considered by a most positive experience at the local level.

Curitiba, Brazil: a reference for urban environmental management

Curitiba, the capital of the State of Parana in South-eastern Brazil, has faced explosive growth since the 1950s. It has changed from a town of 300,000 in 1950 to a metropolis of 2.2 million in 1990, making it Brazil's fastest growing metropolitan area. Initially, various trends indicated a tendency towards uncontrolled growth and the well known social and environmental negative consequences of rapid urbanization. The city's

26

geophysical configuration suggested a physical growth pattern following that of São Paulo, the biggest mega-city in the region, located 250 miles north of Curitiba. The combination of the city's rapid economic surge from a small center for processing agricultural products to a large regional industrial and commercial pole and the mechanization of plantations attracted massive waves of migrants from the countryside. The poverty and income profile in Curitiba is similar to other cities in South-eastern Brazil.

However, even though all trends pointed in the direction of chaotic urbanization, Curitiba has shown that tendencies do not necessarily have to become an ultimate fate. The city has frequently been recognized by the international media, experts and development institutions as a successful example of urban environmental management. From the early 1970s to date, Curitiba has embarked on a different path, following conscious political decisions backed by popular participation. This section examines how this path was laid out, as well as what was learned along the way. Curitiba has had - and still has - a variety of urban-related problems, like any other city worldwide. How did the city administration make a difference? I will try to answer this question by giving an account of three basic aspects: principles, procedures and lessons.

Principles

Priority to people and public transport The city administration consciously decided to take control of the urban growth process. The authorities emphasized growth should take place along prescribed structural axes, allowing the city to spread out while developing public transport that kept shops, workplaces and homes readily accessible to one another. Curitiba's road network and public transport system are probably the most influential elements accounting for the present shape of the city. Total priority was given to public transport throughout the entire city, and to pedestrians downtown.

The city had been growing in expanding concentric circles from the city center. In the mid-1970s, however, the city authorities began to implement the urban design structure that counteracted the unplanned urban sprawl and emphasized linear growth along five pre-determined structural axes. Land use legislation was enacted to guide this growth, allowing for higher housing densities in streets served by public transport. Over the years, urban growth has been encouraged along the structural axes, also known as structural sectors. Each axis was designed as a "trinary road system": the central road has two restricted lanes in the middle for express buses, being flanked by two local roads. There are

27

high capacity one-way streets into and out of the central city one block on either side of this central road. In the areas adjacent to each axis, land use legislation has encouraged high density occupation, together with services and commerce.

The city increased these spatial changes with a bus-based public transportation system designed for convenience and speed. The five express bus lanes existing along the structural axes were complemented by interdistrict and feeder buses that expanded as the city grew. There are large bus terminals at the end of the five express busways, where people can transfer to interdistrict, feeder or intercity buses. Medium-sized terminals are located every two kilometers along the express routes, and a single fare allows passengers to transfer from the express routes to interdistrict and local buses. In high demand routes, tubular, subway-style boarding stations speed boarding times through pre-payment and level boarding. This system replicates some of the advantages of a subway system at the surface, costing approximately 200 times less than a conventional subway. Articulated and the only bi-articulated buses in the world double and triple the capacity of the express busways. [9]

Curitiba has over 500,000 private cars (more per capita that any Brazilian city except Brasilia). Remarkably, 75 per cent of all commuters - that is, more than 1.3 million passengers per day - take the bus. This has resulted in fuel consumption rates that are 25 per cent lower than those prevailing in comparable Brazilian cities, which factor has contributed to the city having one of the country's lowest rates of ambient air pollution. Finally, the average Curitiba resident spends only about 10 per cent of income on transport, which is a relatively low percentage in Brazil.

Designing with nature Flooding was one of the most serious problems Curitiba faced. The city center used to have frequent floods that were worsened by the construction of houses and other structures along streams and river basins. In addition, during the 1950s and 1960s many streams were covered and converted into artificial underground canals that made drainage even more difficult. Necessary drainage works had to be dug underground at a very high cost. At the same time, new developments on the periphery of the city were being undertaken without proper attention to drainage.

Beginning in the late 1960s, some strips of land for drainage have been set aside and certain low-lying areas have been put off-limits for building purposes. In 1975, the remaining natural drainage system was protected by a restrictive legislation. River basins were classified as special areas requiring protection and management, often through park development. Stream protection strips were developed as linear parks and supported by comprehensive tree planting. Other areas prone to flooding were transformed into parks and enhanced with sports and leisure facilities. The parks are also well-integrated in the transportation system via both free green-colored public buses and bicycle paths.

There were several advantages to this "design with nature" strategy. The preventive measures allowed the city to forego substantial new investments in flood control. In 1970, Curitiba averaged only 0.5 m^2 of serviced green space per capita; this figure has now increased one hundredfold to 50 m^2 per person, and all that during a period of rapid population growth. The manner by which this was accomplished is a lesson in environmental management: solving several problems with "win-win" solutions.

Appropriate rather than high-tech solutions Curitiba could have chosen a number of technologically sophisticated solutions to its woes. Two examples illustrate this point. The conventional wisdom was that cities with over a million people needed a subway system to deal with congestion. The other prevailing dogma was that cities that generated over one million tons of solid waste annually required expensive mechanical garbage separation plants. Instead, Curitiba chose different paths for its transportation and garbage problems, based on the principles of simplicity and resource conservation.

The choice of transportation technology was the result of simple economics: an underground metro system would have cost US$ 90-100 million per kilometer, while the express busway system came in at US$ 200,000 per kilometer. Bus operation and maintenance were also familiar technologies that could be operated by the private sector.

Concerning trash generation and collection, Curitiba instituted two innovative programs. The "Garbage that is not Garbage" initiative involves curbside collection and disposal of recyclable materials that have been sorted by households. The "Garbage Purchase" program, designed specifically for low-income areas, seeks to clean up sites that are difficult for the conventional waste management system to serve by exchanging garbage bags collected by residents for bus tokens, parcels of surplus food, and children's school notebooks. Another initiative, "All Clean", temporarily hires retired and unemployed people to clean up specific areas of the city where litter has accumulated.

The results of these challenges to conventional wisdom have been very positive. In addition to the benefits of the abovementioned bus system, the city has today a self-financing public transportation system, instead of being saddled by debt to pay for the construction and operating subsidies of a subway system. The savings have been invested in other priority areas. Concerning solid waste management, over 70 per cent of households participate in the recycling programs. Nearly 1,200 trees are "saved" each day by the volume of recycled paper alone. Sixty neighborhoods with 31,000 families have benefited from the garbage purchase program by receiving nearly a million bus tokens, 1,200 tons of surplus food, and school notebooks in exchange for collecting over 11,000 tons of garbage. These innovations have reduced the costs and increased the effectiveness of the city's solid waste management system while conserving resources, beautifying the city, and providing employment - another "win-win" and low-tech solution.

Innovation and participation The city planners of Curitiba have learned that good systems and incentives are better than just theoretical good plans. The city's Master Plan helped to forge a vision and strategic principles to guide future developments. This vision was transformed into reality by reliance on the right systems and incentives, not on the slavish implementation of a static plan.

An example of a system that yields desirable results is the provision of public information about land. Curitiba's City Hall can deliver information to any citizen in five minutes about the building potential of any plot in the city. Updating the system is a requirement: anyone wishing to obtain or renew a business permit must provide the City Hall with information used to project traffic generation figures, infrastructure needs, parking requirements, and other impacts. The transparent information system helps to avoid land speculation and has been essential for budgetary purposes as property tax is the main source of revenue.

Incentives have also been important in reinforcing positive behavior. Within the city's historic district, owners have the right to transfer the building potential of their plots to another area of the city. This means that historical buildings are preserved and owners are compensated. Businesses throughout the city can "buy" up to two extra floors beyond the legal limit in specified areas. Payment can be in the form of cash or land, which the city then uses as resources for low-income housing.

Procedures

Time is money The longer it takes to implement solutions, the more expensive they become. Cities are not static and nor are solutions. For example, a low-cost sanitation technology that is suitable for a density of 40 families per hectare will not be suitable for a density of 100 families per hectare when the population grows, and a more costly approach might be required. Curitiba has developed alternative approaches to deal with the pollution of Iguazu river tributaries. For certain areas, however, demographic density still demands conventional approaches which represent a heavy burden on the municipality's budget. The same applies to urban design, public transport technologies, waste management techniques, low-cost drainage approaches and urban services in general.

Prevention vs. remediation With regard to environmental and urban infrastructure, for instance, it is well known that the cost of prevention may relate to the cost of remediation by a factor of 1 to 100. This applies to transport, sanitation, waste management and various other issues. In other words, it makes sense to spend one dollar today in order not to have to spend one hundred dollars tomorrow. The planning of the "structural sectors" in Curitiba is an example that demonstrates how transport costs were saved directly. Indirectly, there were savings in infrastructure improvements such as water, sewage, electricity and communication. Regretfully, most cities consider urban development interventions only when it is too late or too expensive to prevent problems. Apart from the world's mega-cities, there are also thousands of developing municipalities which do not necessarily have to make the same mistakes.

Insistence pays People are sensitive to any city administration that shows signs of constant care. Maintenance is of paramount importance in providing the population with an indication of how much the city administration is concerned. A simple example: Curitiba decided to regularly plant flowers on what had been the main avenue of the city center, and was then made the first pedestrian street in Brazil. At the time, people were not used to seeing flowers on the streets, and they were often picked up or vandalized. With time, insistence and a regular maintenance scheme, the population began to respect, and even defend, the flowers. I have personally witnessed, on various occasions, people confronting those who had picked up a single flower. The city's culture is permeable to administrative care.

Institutional streamlining How will the decision-making process work in City Hall, where there are thousands of employees and hundreds of bosses? In Curitiba, three different functions were developed separately, though with constant inter-facing, namely: planning, execution and administration. Weekly meetings between the Mayor and the key actors in each one of these areas defined the targets for the week. The execution of these targets was closely monitored during the following weeks. The streamlining of institutional actors is as important as the integrated

31

participation of personal actors. Usually, participation can be conflictive or consensual, but it should ultimately lead to a sensible balance between planning, execution and administrative functions, in response to demands from the population.

Incremental learning The perfect plan will never be implemented. Rather than pursuing perfection, Curitiba concretely did what was possible to do at specific moments in time - and incrementally developed such ideas in practice when they were already operational. The land use legislation, the industrial city, bicycle paths, the parks policy, the bus design and the design of integrated public transport terminals are examples of this approach. It should be always clear that an "idea" has three components: the idea itself, its feasibility and its operation. This sequence forms a circle, which then leads to further improvements. Curitiba's initial Master Plan is the first basic example: a book prepared in 1965 with many good potential ideas, which were later changed or improved by the first administration of Mayor Jaime Lerner in 1971 to be viable and operational.

Participatory field work precedes desk top design One advantage of working for local administrations is that the issues are concrete and the problems are just outside your door. This level of clarity is not easily attainable in broader levels of administration, such as States or countries. There is always a bureaucratic way of dealing with any issue and this is certainly the best way not to solve it. Planning officials, architects and other professionals in Curitiba have always been encouraged to look at the problems, talk to the people, discuss the main issues, and only then reach for the pen. This behavior has been exercised by Mayor Lerner and other mayors in Curitiba. As far as politics are concerned, people normally know when a politician is on their street just to collect votes or to collect concrete suggestions. A genuine concern in looking at the problems and talking to the people, at any level of decision-making, provides a new insight which is seldom self-evident at the drawing table.

Possible lessons for an urbanizing world

Some of the lessons from the Curitiba experience which other cities could use include:

1　Top priority should be given to public transportation rather than to private cars, and to pedestrians rather than to motorized vehicles. Bicycle paths and pedestrian areas should be an integrated part of the road network and public transportation system. In Curitiba, less attention to meeting the needs of private motorized traffic has generated less use of cars.

2 There can be an integrated and environmentally-sensitive action plan for each set of problems, but solutions within any city are not specific and isolated, but interconnected.

3 The action plan should involve partnerships between responsible actors such as private sector entrepreneurs, NGOs, municipal agencies, utilities, neighborhood associations, community groups, and individuals.

4 Creativity can substitute for financial resources. Ideally, cities should turn what are traditional sources of problems into resources. For example, public transport, urban solid waste, and unemployment are traditionally listed as problems, but they have the potential to become generators of new resources and solutions.

5 Creative and labor-intensive ideas can, to some extent, substitute for capital-intensive technologies. Cities do not need to wait for bailouts or structural reforms to begin working on some of their problems.

Other lessons include:

1 Even during a period of rapid demographic growth, physical expansion can be guided through integrated road planning, investment in public transportation and the enforcement of appropriate land use legislation.

2 Technological solutions and standards for everything from public transit to recycling should be chosen on the basis of affordability.

3 Integrated solutions can be implemented through partnerships between key actors. This often requires that the network of formal and informal economic relations be supported - and not hindered - by urban managers.

4 Public information and awareness are essential. The better citizens know their city, the better they treat it.

These principles may seem like simple doses of common sense, but they have rarely been applied to ailing cities around the world. Perhaps the missing ingredient is political commitment and continuity: Curitiba's leaders have pursued their common-sensical path for over two decades. Beyond the city, the Curitiba case suggests that State and national governments would do well to acknowledge the strategic importance of cities as potential instruments for positive development and change.

These lessons are being learned by other cities, inside and outside of Brazil. In Brazil and other Latin American cities, pedestrian walkways, bus lanes and waste management programs that were pioneered in Curitiba have become popular urban fixtures and procedures. Cities in regions as different as Africa, Asia, North America and Europe have expressed interest in the approaches put to practice in Curitiba. Naturally, one size does not fit all, solutions need to be tailor-made, adapting to local circumstances. Not all cities enjoy Curitiba's political will and continuity. However, at least one of Curitiba's many creative, resource-conserving solutions may fit many cities that make up an increasingly urban world.

The challenge of replicability

Cities are not ideal subjects for scientific analysis. Urban development is the more scientific of arts, or the most artistic of sciences. The traditional scientific method, in very generic terms, is based on the principle that certain physical phenomena can be repeated in the same way if basic conditions are in place. Therefore, based on the expected results, scientific laws are established. Science, based exclusively on "experience", can not accommodate the diversity and chaotic behavior existing within each city. Replicability laws cannot be established for urban development: common sense and hard work remain as basic rules.

After having presented the case of Curitiba on various occasions, the most frequent questions are: why are there not more "Curitibas"? What made Curitiba successful? How to transfer the positive lessons to other cities? The truth is that the experience of Curitiba only became well known some twenty years after its beginning. It has been a consistent, progressive and incremental process. Mistakes were made and later corrected. Curitiba still has serious sanitation, education, and housing problems to tackle. The challenge now is to continue the successful policies and programs, while applying these lessons to unresolved issues of urban development.

With regard to its replicability, two types of experience will be briefly described: contacts between Curitiba and other cities in Brazil or other countries, and the more recent extension of the experience of Curitiba to other cities in the State of Parana. As far as contacts with other cities are concerned, Curitiba has attracted considerable attention and became a destiny for various technical teams coming from other parts of Brazil and abroad. Among other tasks, I have been personally responsible for meeting professionals, mayors and council persons from other municipalities and for discussing with them the problems faced and lessons learned by Curitiba. On average, there have been two to three visits in a week. A few observations have emerged from this process:

1 While some visitors refer to Curitiba as a "model", others refer to the city as an "exception". I would argue that it is neither. No city is so exceptional that some basic working principles cannot be analyzed and transferred to other cities; nevertheless, cities are intrinsically different, and it would be a mistake to choose a "cookie cutter" approach, no matter how successful a specific experience has been.

2 The different opinions expressed above probably depend on the extent to which the situation in other cities is similar or not to that of Curitiba. It has been possible to infer that, while some cities apprehended lessons due to the similarity of their situation and problems with Curitiba's, some other cities seem to have grasped inspiration by contrast: "our city is different, and now that we have seen this solution working in Curitiba, we know we have to somehow adapt its elements to our own context."

3 Representatives from many other cities visiting Curitiba ask about a possible "implementation formula" as to what made the solutions in Curitiba work. I would outline ten basic elements, namely: political will; leadership; continuity; community involvement; courage to develop incremental solutions (including "trial and error"); informality and commitment of the technical team; search for simple, appropriate and affordable solutions; team spirit; hard work; and good timing. These elements do not represent in themselves a formula, nor a pre-requisite. However, they have been present in varying degrees throughout the various municipal administrations during the past thirty years.

Other cities would also regularly ask for a flowchart, some kind of scheme showing boxes, circles and arrows, indicating how the various institutional and social actors related one to the other during the planning, implementation and evaluation of projects. It may be relatively easy to represent reality through diagrams, but these are of very limited value in reflecting the comprehensive and informal intricacies of urban governance and in transferring experience. It would be pretentious and wrong to represent complex processes such as urban development through caricatures. Perhaps due to these reasons, no flowchart was developed to explain the entire experience of Curitiba.

Many questions are usually asked about successful transfers, "twinning", technical cooperation agreements, or simply the extent to which the experience of Curitiba has been "exported" to other places. Brazilian cities such as Rio de Janeiro, Sao Paulo, Vitoria, Joao Pessoa, Santos, Niteroi, Florianopolis, Rio Branco, Belo Horizonte and Porto Alegre have expressed varying degrees of interest in Curitiba, as well as foreign cities such as New York, Cape Town, Montreal, Toronto, Quito, Himeji, Lagos, Caracas, Santiago, Dakar, Shanghai, among many others. The city of Curitiba does not claim the merit of having changed urban planning practices in other cities. Nevertheless, it is fair to say that, at least in the Latin American context, certain initiatives were pioneered by Curitiba and possibly inspired other practices. A few examples: the transformation of the main downtown avenue in a pedestrian mall; priority for public transport; implementation of segregated bus lanes; development of household-based waste management schemes; development of "garbage exchange" for bus tokens, food and school notebooks; incremental land use development in association with a hierarchy for the road network.

Curitiba is the capital of the State of Parana, which has a total of 339 municipalities, a population of 8,5 million and an approximate budget of around US\$ 3,5 billion. In January 1994, former Mayor Jaime Lerner was elected State Governor, being today in a position to extend the experience of three successful terms of office in Curitiba to the whole State of Parana. The following are some of the main initiatives being undertaken to promote rural-urban linkages.

Rural Villages

The "Rural Villages Program" is an initiative of the state of Parana aimed at curbing rural migration by providing housing, infrastructure, services and jobs to rural and peri-urban dwellers. It has been partially inspired by a previous "rurban community" project implemented in Curitiba, now extended to the entire State. The rural villages are formed by small properties of 5,000 m^2, where the rural worker lives and develops agriculture to support his family and sells the surplus. Each rural village has less than 200 properties, which are purchased by dwellers through an affordable 25-year mortgage program. The area is provided by the participating municipalities. Families have 30 months to begin paying for the land through monthly installments that do not exceed 20 per cent of their small incomes.

The rural villages are always implemented near secondary roads, making the access to schools, health centers and basic shopping/trade easier. Municipalities do not have to invest too much in infrastructure to serve these communities. The module house has 45 m^2 and is built through a self-help scheme by their future owners. Potential dwellers have to apply to the State and are selected by the following criteria: the head of household should not be older than 55 years; the head of household should have lived in the municipality for at least four years; families who already own property cannot apply; the family should have an income between one and three minimum monthly wages (one monthly minimum wage in Brazil is equivalent to about US$ 100); and families should preferably have underage children, while the head of household should exercise a temporary job in rural or peri-urban areas of the municipality.

The Rural Villages Program is supported by a partnership between various governmental institutions. While the municipalities involved provide the land, the State Secretariat for Agriculture and Food approves plans for new communities according to land exploitation potential as well as technical assistance and training for a period of 30 months. The State Secretariat for Housing Policy approves plans for new communities in cooperation with the above, provides financial resources through direct loans to dwellers and provides guidance and technical assistance for housing construction. The State Secretariat for Children and Family Affairs supports newly established families through social family monitoring, as well as being in charge of the promotion of technical and income-generation training courses after the completion of houses and communities. The State Secretariat for Planning is responsible for the overall planning and coordination, while the State Secretariat for the Environment is in charge of enacting and enforcing environmental legislation and collecting and mapping land and natural resource information. Finally, the State Secretariat for Employment and Job Relations is responsible for the follow-up of working relationships and job generation aspects during implementation, the State Company for Electricity for the electricity infrastructure, and the State Company for Water and Sanitation for the water and sanitation infrastructure.

The State of Parana aims to assist 60,000 families through this project. As of August 1996 some seventy rural villages were operational or being implemented throughout the entire state. While it is rather soon to undertake a comprehensive evaluation of the program, it can be said that it definitely represents in principle a creative and participatory alternative to conventional agrarian reform practices.

The Rural Roads Program

Rural roads are fundamental to promote rural-rural and rural-urban integration in the countryside of the State of Parana. They have a role in promoting transport, commerce and access to schools and health centers, enhancing economic and social contacts which are essential to promote sustainable development in rural areas. The improvement of transport and communications can help avoid rural migration, as dwellers have more direct and frequent access to urban and peri-urban centers, to agriculture storage and exchange facilities, without having to abandon their rural roots.

The Parana State's road network is comprised of federal, State and municipal roads. Federal roads total some 3,300 kilometers; State roads comprise some 12,400 kilometers and municipal roads some 245,000 kilometers. Municipal roads are precisely those which ultimately reach the sources of agricultural production; however, only one per cent of all municipal roads are properly paved. The objective of the Rural Roads Program is to facilitate permanent basic traffic conditions in rural roads between agricultural production areas and storage and commercialization facilities and towns, thus benefiting small-scale producers and supporting access of low-income rural populations to basic health and education services.

Given the immense extension of the existing rural roads network vis-à-vis available financial resources, technical characteristics were kept as simple as possible. This means throwing away standard road construction manuals and avoiding higher costs normally charged by conventional contractors. The basic objective is to execute a 6,40 meter wide flat platform, with a single central paved lane of 3,60 ms. Paving materials vary according to local availability, the technology is labor-based and not capital-intensive. Low-cost drainage facilities are kept to a minimum.

Five pilot rural roads are being developed following this approach and the results so far seem satisfactory. Each road required a different technical solution, depending on their geo-physical situation/topography, availability of construction materials and maintenance needs. Different companies have been involved with each project, following competitive bidding. The average cost per kilometer has been approximately US$ 40,000. The total target is the development of approximately 1,000 kilometers of rural roads.

The "Teachers' University"

A "new town" has been developed close to the city of Faxinal do Ceu in the countryside of Parana, to support seven-day advanced education workshops aimed at the 80,000 public teachers of the State - comprising elementary, high school and university levels. Accommodation, restaurant, meeting rooms, leisure facilities and all necessary

38

infrastructure have been built to support these "educational jamborees" designed to get teachers to share their experiences through special retreats. During these intensive workshops, teachers participate in daily journeys that can begin at 7:00 a.m. and end at 11:00 p.m. Sessions on subjects are as varied as Eastern Thinking, Opera, Fitness, Dancing, Painting, Philosophy and Psychology. The courses expand the knowledge of those who are forming the future generations.

Every week, 144 working groups follow the seven-day workshop. Each working group cannot exceed seven participants. All working groups also follow a three-phase program from Monday to Thursday. In the first phase, they establish the group's internal management rules and undertake their own assessment of the main problems faced by teachers in their respective areas of work; the second phase focuses on possible solutions for the their problems, and the third phase is dedicated to the development of a report emphasizing successful concrete teaching experiences. Out of 144 reports, 24 are chosen every week to be presented to all groups in the last day of the seminar. By the end of 1996, there were some 1,000 reports in a database to be circulated around the entire State.

The "Teachers' University" counts on 12 full time resident staff members and a total of 24 inter-disciplinary professionals forming the management team. The seminars are also supported by 48 supervisors, who help link participants with coordination staff. Some of the most experienced professionals in the State are invited as guest lecturers to speak about issues as diverse as nature conservation, poetry and politics in Ancient Rome.

Program for the Indigenous peoples of Parana

The State of Parana was originally occupied by various Indian groups, of which the Kaingang and Guarani survive. The Xeta tribe, for instance, has been approached by "civilization" in the 1950s and are practically extinct; only six individuals remain. The Indigenous communities currently grow at four per cent a year, a relatively high growth rate if compared to the average 1.5 per cent observed in the South of Brazil. However, due to lack of choice and opportunities, their reduced members are constantly migrating to cities, where they increase homelessness and joblessness statistics. Although their culture and traditions have contributed to preserve some of the most extensive pinewood forests in the State, their land is constantly threatened by the "white men", mostly loggers, farmers and land speculators.

The aim of the abovementioned program is to preserve their culture, language, traditions and land, while at the same time trying to revert the problem of poverty (and hunger) which affects most Indigenous communities in Parana. The first phase focuses on the Mangueirinha reserve, having two basic priorities: support the social, cultural, economic, environmental sustainability of the Indian community; and preserve the largest *araucaria angustifolia* (a native pine tree) forest worldwide. The Mangueirinha reserve has a total area of 17 million hectares and a population of 1,617 Indians. The State Secretariat for the Environment is undertaking the following initiatives: provision of latrines, septic tanks and other basic sanitation facilities; alleviation of the impact caused by roads crossing the area through signing, "traffic calming" road works and incentives to tourism; the existence of the Indian reservation is being clearly marked by gateways and specific points for the sale of handicrafts are being created; and special education programs are being developed for the 300 Indian children attending school. Moreover, new houses are being built taking into consideration the lifestyle of Indian communities and locally available building materials; specific monitoring points and control towers are being created to enforce the legal exploitation of natural resources in the area; and social monitoring and support to Indian communities is being provided.

Apart from the above, an environmental education initiative is being developed, based on a successful program for children in low-income neighborhoods already implemented in Curitiba. This initiative aims to preserve Indian traditions and to provide job training. Indians are receiving training on how to raise cattle; technical assistance, basic equipment and seeds for the development of fruit crops; construction of small dams for pisciculture; construction of a brick factory using locally available clay as raw material; training for sustainable forestry practices; training for the operation and maintenance of basic agriculture machinery; and training for the extraction and sale of *pinhao* (edible pine tree seeds). In each village, the government offers a food subsidy during six months for each family during the first year of the program.

Replicability lessons from the State of Parana

The experience of Curitiba has developed into a useful reference for other cities in Brazil and worldwide. However, urban-related problems do not exist in isolation from the surrounding countryside, and the extension of Curitiba's lessons to other cities in the State of Parana already provides some elements for the analysis of rural-urban linkages and the replicability of urban governance experiences.

As was mentioned above, following three alternate terms as mayor of Curitiba, Jaime Lerner was elected Governor of the State of Parana in January 1994. As Governor, former Mayor of the capital, recipient of various international awards and holder of high approval ratings among voters and strong credibility, Governor Lerner is today in a position to influence municipal policies across the State and to propose a series of concrete interventions inspired by the experience of Curitiba.

The distribution of human and financial resources from the State government to municipalities is not being subject to political criteria; for example, according to all the information received, all mayors receive equal support regardless of their political parties. Technical assistance is not developed in a patronizing manner. Mayors of recipient municipalities have to contribute resources as well as the people who benefit from the programs.

The activities being developed clearly indicate the policy priorities chosen to promote rural-urban linkages. Job generation for the provision of basic needs, housing/services, transport and education are the pervasive themes existing in the main programs briefly described above. The approaches being developed so far also indicate that potentially successful sub-regional programs do not depend on the availability of massive financial resources. Proper technology choices, participatory approaches, informal solutions and clear institutional roles can help promote affordable initiatives.

Promoting rural development without causing further rural migration is challenging. The poor are not satisfied to remain poor: quite often, as soon a family's income increase and they accumulate savings, migration of the head of the household to the nearest town or city may follow. Many rural development programs have failed as they ended up by increasing rural migration. The approaches being adopted in Parana may avoid this tendency due to the following characteristics:

1 Participatory programs are instilling a long-term sense of responsibility, thus becoming sustainable. In the rural villages, for example, instead of receiving a subsidized home, families receive basic assistance and a loan to buy the land and build their own home. Families have to pay for the land, on which they choose to develop an activity that they are familiar with. Rural migrants normally have to abandon their knowledge of agriculture to survive in a city. By facilitating access to their own land, the program motivates families to continue the activities they were already developing on other people's land, without creating any significant disruption in family life.

2 Complementary education programs contribute to avoid rural migration. No family voluntarily migrates if they see their children having access to public school, receiving free meals and standard education. Once motivated, teachers improve the quality of education and children attend school more often. The existence of training courses for adults also prevent the frequent lure for better education opportunities that normally exist in cities.

3 The attention being given to transport, beginning with rural roads, is an important element. By háving access to storage facilities and commerce/trade points, rural producers are motivated to remain on their land, while enjoying access to commercial channels normally associated with urban areas. Schools and health facilities also become more accessible.

4 The technical team now in charge of the State of Parana is similar to the technical team that was in charge of Curitiba. Many former Municipal Secretaries were appointed State Secretaries by Governor Lerner. On the one hand, it is a competent, experienced and motivated team. On the other hand, this team is facing the added challenge of having to execute projects through existing municipalities. The playing field is not even: the technical capacity of municipalities in Brazil can vary tremendously from one area to another. It is still too soon to determine to what extent Governor Lerner and his team are carrying out their work with other municipalities in a homogeneous manner. Nevertheless, the creative approaches being adopted and the initial results are inspirational and encouraging.

Conclusion

The main conclusion from the abovementioned experience is that to think urban is indeed to think hope, and not despair. Cities may not be intrinsically sustainable, but they can definitely be made more sustainable and contribute to people-centered sustainability. The urban environment permits the efficient mobilization of people's energies and human potential, as it enhances the income-generating ability of the informal sector and promotes education and skills training. Cities spearhead economic development, and they transform society through growth in the productivity of labor. The economies of scale found in cities make them capable of generating goods and services in excess if their share of the national population.

The current international scenario seems to favor enhanced relationships between the international community, national governments and local authorities. The potential of cities to support a path towards sustainability gradually becomes acknowledged by national governments. At the same time, considerable work needs to be undertaken before the cities and peri-urban areas actually change positively for all their inhabitants. Approaches that regard urban areas as a continuity of peri-urban and rural areas need to be included in the policy agenda.

Cities produce a large part of the Gross Domestic Product, indicating a high per capita productivity. Markets for labor, capital and technology transform advantages of location into higher incomes and employment opportunities. Urbanization restructures social relations and improves life expectations; child-bearing women usually insist on smaller families and are introduced to family planning notions when moving to urban and peri-urban areas.

Greater participation in the urban economy leads to the creation of a skilled and literate work force. In short, the full use of the human development potential in urban areas is directly related to the promotion of policies conducive to improvements in the urban environment and urban economics.

Notes

1　See United Nations Population Division (1995).
2　See United Nations Development Program (1992).
3　See United Nations Development Program (1994).
4　See United Nations Development Program (1995).
5　See Rabinovitch (1996).
6　See United Nations Development Program (1996).
7　The International Union of Local Authorities-IULA, the United Towns Organization-UTO, Metropolis and the Summit of the World's Major Cities formed the "Group of Four" at the time of the International Colloquium of Mayors for Social Development, held at the United Nations in August 1994. Other global and regional cities' associations later joined, forming the so-called "Group of Four Plus", which participated under this name at the Second United Nations Conference on Human Settlements (Habitat II).
8　*"Isto E"*, Brazilian magazine, No. 1,404, August 1996, Sao Paulo, Brazil.
9　See United Nations Development Program (1993).

References

Rabinovitch, J. (1996) *Second United Nations Conference on Human Settlements/Habitat II UNDP Report*, UNDP unpublished report: New York.

United Nations Development Program (1992) *Cities, People and Poverty - Urban Development Cooperation for the 1990s*, UNDP: New York.

United Nations Development Program (1993) *UNDP's Urban Development Cooperation*, UNDP booklet: New York.

United Nations Development Program (1994) *Human Development Report 1994*, Oxford University Press: New York.

United Nations Development Program (1995) *UNDP Strategy and Action Package for Habitat II and Implications for Follow-up Activities*, UNDP working document: New York.

United Nations Development Program (1996) *Decentralized Governance Program (draft)*, UNDP/Management Development and Governance Division: New York.

United Nations Population Division (1995) *World Urbanization Prospects - 1994 Revision*, UN: New York.

3 The role of NGOs and CBOs in a sustainable development strategy for Metropolitan Cape Town, South Africa

Vuyiswa Tindleni*

Introduction

The Cape Town Metropolitan Area (CMA) is surrounded by natural barriers, the Atlantic Ocean to the West and South, False Bay to the South-east and Hottentots Holland evolution to the North and East. It is a sprawling city which is home to a population of about 3.6 million people. The CMA is expanding rapidly and has large and growing peripheral urban settlements. Cape Town has a Mediterranean climate, and at 34 degrees it is far enough south to jut into the West wind drift in the winter. Cape Town is exposed to cold wet weather from the North and West in winter and strong Southerly and Easterly winds in summer.

South Africa is prominent in its diverse culture, language and environment. In climate, natural environment and majority politics Cape Town is different, but with respect to diversities of people and sharp disparities between rich and poor in access to services it is similar to much of the rest of the country. Cape Town is not a peaceful city: it has a high rate of petty and violent crime, and is at present experiencing a show-down between organized crime and vigilante groups. The Cape Town City Council, recently renamed the Central Substructure, is the largest of six sub-structures which constitute the CMA. It is in the process of revising policies and strategies to address past disadvantages and discrepancies which characterized the past.

The Coalition for Sustainable Cities (CSC) is making an effort to work in partnership with the Cape Metropolitan Council and the Central

Substructure to confront urban developments needs, as a requirement of environmental justice in a new democratic dispensation. The CSC is a working group of a non-governmental network known as the Green Coalition, which has a membership of 53 environmental organizations, services organizations and trade unions. The Green Coalition is a regional branch of the national Environmental Justice Networking Forum. One of the main objectives of the CSC is to establish a strong proactive platform on which to make a sustainable city of Cape Town, by addressing the need for increasing social equity, environmental sustainability and economic efficiency.

The Cape Metropolitan Council regards the involvement of the CSC as an opportunity for proactively addressing environmental management needs. The new order in South Africa, enhanced by influences such as those espoused by the Local Agenda 21 and the International Council for Local Environmental Initiatives (ICLEI), have created favorable conditions for the establishment of such partnerships between local government and civil society.

This paper is a compilation of inputs of CSC member organizations, and addresses many of the issues and processes currently underway in the movement towards a more sustainable development in the CMA.

The formation of Reconstruction Development Program forums in South Africa

A feature of South Africa has been the combination of racism, poverty and repressive labor practices whilst the concentration of wealth in the hands of a minority has resulted in the over-utilization of non-renewable resources, factors which have contributed to significant environmental degradation. This situation, which is mirrored by many examples within cities, countries and between the North and South, has created a major challenge to the environmental sustainability of the CMA, the Western Cape region, and the South African nation as a whole.

South Africa expects the Reconstruction and Development Program (RDP) to be a key resource for addressing the need for a people-centered environment. All political parties in the National Assembly have committed themselves to the RDP's main objective, which is to create a platform through which all relevant stakeholders are brought together with the primary purpose of promoting ongoing participation, and involvement, in the management and implementation of the Program through local forums. The ultimate aim is for communities to take control of their own lives.

The role of the forums is to ensure more integrated and sustainable development especially through constant and widespread consultation and constructive interaction with government structures, so that representative governance is eventually achieved at the local level. The CSC's intention is to increase the ability to restore environmental equilibrium and to reduce the negative effects of past discrepancies.

Local authorities are the key institutions for delivering basic services, extending local control and preserving natural environments. To achieve this goal, emerging democratic local authorities must work in partnership with Community-Based Organizations (CBOs) and Non-Governmental Organizations (NGOs) to establish minimum conditions of good governance and to implement effective development projects. Activities in this direction are advancing, as addressed below.

Land use and transport

The Metropolitan Spatial Development Framework (MSDF) originated from the need to redress, both spatially and a-spatially, a historical legacy of poverty, inequality and under-development. It seeks to provide the basis for controlling urban growth, development and change. The MSDF aims to ensure the maintenance and enhancement of the CMA's natural, built and cultural resources, with a view to establishing long-term sustainability. In essence, it is a negotiated framework in which to address land-use issues within the CMA. Although both the MSDF's mandate and the political will for its implementation require redress, it has historic relevance with respect to the breadth of issues it addresses.

The mission of metropolitan planning

The core point of the MSDF's mission is that the metropolitan region is a "gold mine", in that it is located within an environment that is globally unique and should be conserved and maintained on a sustainable basis to serve as the central component of the region's economic base. Through the development of tourism and mass recreation leading to job creation, social mobility and economic development, this asset can be further realized.

The goals of planning and development should be the promotion of the following factors: equitable, convenient and affordable access to opportunities; the creation of a vibrant environment, rich in affordable opportunities and choices; economic growth, prosperity and job creation; social well-being; the protection and enhancement of the CMA's uniqueness and beauty; and the creation of an adaptable urban structure. Furthermore, such goals should also promote increasing levels of safety, comfort and confidence; openness and accountability in decision-making; the efficient use of all resources; and the management of natural and built environments to meet the needs of existing and future generations.

According to the CSC, in order to generate the economic resources necessary and to secure the infrastructure needed for a sustainable city, the method of least-cost planning must be adopted. These costs must include those associated with addressing environmental degradation, the potential for job creation, as well as the increasing distribution of resources and other positive and negative externalities that are not considered when growth or Gross Domestic Product are considered as the main indicators of economic success. This approach has been attempted by the Cape Metropolitan Council's Planning Department, which over the last five years has gradually supported utilizing "process principle" planning. Recently, task teams have solicited input and compiled reports on some of these issues, which are requested prior to the negotiation and formulation of final legislation.

Model Communities and the Local Agenda 21

The Central Substructure is a member of the ICLEI. The Model Communities Program (MCP) is a partnership project of the ICLEI, which emerged from the 1992 Rio de Janeiro "Earth Summit". The project is in keeping with Chapter 28 of the Local Agenda, which stated that, by 1996, Local Agenda 21 should be in place through consultation and consensus with local people, business and other organizations. Women and youth should be involved in decision-making. Local authorities and international organizations were mandated to strengthen cooperation and coordination and also the exchange of information between local authorities.

A partnership between the Central Substructure and the environmental sector was explored in a series of large facilitated workshops, the first of which was on 15 July 1995. The intention of the workshops was to introduce the concept of Local Agenda 21 to the CMA. A decision was taken to emphasize the natural role of RDP forums in executing the MCP, since the feeling was that the program would only work if it were implemented within an existing RDP forum.

A second workshop was held on 7 October 1995, in which communities and NGOs were introduced to the MCP. A third workshop which explored partnership in the environmental management of the metropolis was held on 28 October 1995. This workshop established a basis for planning processes for the MCP to monitor environmental injustices. A selection committee was set up to select a candidate for a pilot of the program. A survey to audit needs was circulated in the Western Cape, and the suburb of Hanover Park was selected.

Hanover Park falls within the Central Substructure and met the criteria of needs that had been defined in previous workshops. Meetings were held with the CBOs and NGOs and members of the existing RDP forums to introduce the Local Agenda 21 and the MCP. The main intention was to form a steering group from the partnership between an environmental NGO, Hanover Park RDP and the local government. This steering committee has selected a researcher/development facilitator who will take the process forward with ICLEI, a funder (potentially USAid), and the local partnership.

An integrated approach to sustainable energy

Pressures on the local environment arise both from over-consumption of resources by the wealthier sector of the population as well as the demand for access to resources to meet basic needs from the poor sectors. It is estimated that, of the 500,000 households in the CMA, about 100,000 still do not have access to electricity. The level of access in Cape Town is far higher than the national average of 50 per cent.

The framework of our city's sustainable development path must rest on the following principles: emphasis on access to adequate and affordable energy services; renewable energy use and efficiency encouraged wherever possible; research and piloting of cleaner forms of energy; least-cost energy planning, taking social and environmental externalities into account; decentralization and diversification to increase security of supply; and transparency in order that all may understand and participate in the system.

The CSC takes a particular interest in energy policy piloting, and wants to contribute towards the sustainable development of the CMA in this field. The CSC identified the following key concerns on energy issues: metropolitan energy policy formulation and demonstration; grassroots participation in defining energy problems and solutions; improvements in the energy efficiency of new and existing house stock; urban densification and appropriate transport infrastructures; reduction of the negative respiratory health impacts associated with airborne pollution, both inside and outside the dwelling; and building with suitable designs and material compatible with the climate, so as to reduce the prevalence of cold and damp housing.

Electricity

Eighty per cent of the primary energy in South Africa is derived from coal, most of which is used in the generation of the lowest priced electricity in the world. Cape Town receives its electricity through the national electricity grid operated by the commercialized parastatal, Eskom, that generates 99 per cent of the country's electricity, 94 per cent of which is produced by coal fired power stations, five per cent by the Koeberg nuclear power plant and one per cent by hydroelectric schemes. With the CMA having an above average wind regime, the Council for Scientific and Industrial Research, in a consortium with the Central Substructure's Electricity Department, are in the process of arranging for the installation of two or three old wind turbines donated by Scottish Hydro. Other initiatives to promote wind energy are currently in planning stages, but much depends on the nature of electricity regulation, which is in a state of flux.

The way energy is supplied and used can either contribute or detract from the potential for sustainable development. In the past, control over electricity was also undemocratic. The energy policy focused on the self-sufficiency of the internationally isolated South African state, at enormous costs to the national economy. During the transfer to democracy, it was recognized that energy policy should be developed on the basis of an integrated understanding of demand and supply and should focus on meeting the basic energy needs associated with clinics, schools, and low-income households.

A National Electrification Forum was established in 1994 and was followed by a National Electricity Regulator being charged with the restructuring of the electricity supply industry. Essential in that process is the setting of a national domestic electricity tariff structure, and the rationalization of the number of distributors. Currently, as many as 400 exist, most of which sell electricity on a cost-plus basis. The rate of electrification nationally is nearly 500,000 households a year. Pre-payment metering systems are being installed in most dwellings as part of the electrification campaign. A downfall of an otherwise successful campaign has been the fact that it has almost exclusively aimed at supplying electricity to homes without improving the potential for efficient use. This has precipitated the development of a rapidly increasing peak demand for electricity in the most populous areas in the coldest days of the year. In order to mitigate this trend, Eskom has recently allocated substantial resources to shift load, reduce peak demand and improve end-use efficiency in their demand-side management program.

Paraffin/kerosene

The informal structures existing in the CMA are especially at risk from open flames. In overcrowded conditions, lamps and candles are easily knocked over and hundreds of dwellings, with their meager contents, are destroyed by fire each year. These households often have no other alternative but to use paraffin/kerosene for cooking, space and water heating and lighting. These houses are often not well ventilated and toxic fumes and high humidities build up. This can exacerbate the spread of respiratory diseases particularly amongst women and young children.

In fact, there is an increase in admission of paraffin poisoning at Cape Town's Red Cross Children's Hospital. As paraffin is often bought or stored in soft drink bottles, there is a high incidence of children drinking paraffin, mistaking it for a soft drink. Oil companies have initiated a campaign to introduce childproof lids for the bottles and provide information in the case of ingestion.

Transport

In 1995, the Department of Mineral Energy Affairs stated that urban transport has been characterized by high levels of passenger car ownership and usage. It went on to describe the current urban planning process which assumes an increase in traffic growth rate, and seems to aim at facilitating the flow of traffic to accommodate an increase in passengers within majors centers. Studies undertaken in the 1980s showed that children who lived along Cape Town's major roads had high levels of lead in their bodies, as a result from lead pollution from cars.

Within the CMA, people travel up to 60 km to work on public transport networks which are overcrowded, expensive, unsafe and inadequate. It is becoming increasingly apparent that the solution to the inadequate transport situation is linked to restructuring the form and structure of the CMA. A sustainable transport system for a city of 3.6 million people cannot be based on private automobile ownership. There have been multi-million Rand upgrades of the major motorway into Cape Town, the N2. A bus lane was created, as a pilot project, operating during peak hours. It is open to buses and minibus taxis, but there has been little public education around the project.

51

The minibus taxi industry is a rapidly growing transport industry, which has precipitated many savage conflicts throughout the country, but it is as yet unregulated. After protracted negotiations with the Cape Town City Council, the city now has a taxi rank policy which is adequate for our current need.

Grassroots capacity building policy and governance

In 1995, the Energy and Development Group based in Cape Town held a number of participative workshops around national energy policy development. Using innovative methodologies, grassroots communities were able to trace the chain of energy usage, to pinpoint inefficiencies or problems they experienced, and to suggest recommendations which were later assessed as energy policy options. The results showed that policy processes can be made simple, transparent and participative leading to a greater accountability and confidence in taking issues to recipient institutions, notably the National Energy Policy Summit.

In November 1995, the Science Policy Advice Unit held a workshop in Cape Town to inform NGOs, civics and unions on all aspects of the nuclear industry. The intention was to prepare members of civil society for the National Energy Policy Summit, empowering them to formulate an informed position with respect to the nuclear industry. In the past, the apartheid government threw a curtain of secrecy around their nuclear program, which then included the Koeberg nuclear plant only 30 km from Cape Town, and the manufacture of six atomic bombs and nuclear waste storage sites located near poor, ill-informed communities. The anti-nuclear campaign and the CSC aim to raise awareness by lobbying parliament and are calling for an external evaluation of the entire nuclear industry.

The gap between rich and poor is not only in access to resources such as adequate shelter and public transport, but also in access to information. Issues of housing and energy efficiency are on the agenda of policy-makers, but communities on the ground lack sufficient information to make informed choices or to lobby for renewable energy supply.

An example of this situation is Delft, which is a low-cost housing estate where cheap and poor workmanship and material led to the deterioration of the houses. Many people died when houses rushed up prior to the 1994 elections collapsed in winter storms less than six months after they were built. In addition, houses without proper ceilings exacerbated respiratory illnesses, particularly amongst children. A partnership between CSC and Delft residents was formed, called COLD

(Coalition for Life in Delft), proposing the concept that houses could be built in the region that require no space heating in winter. Soon afterwards a presidential task group took over the investigations, but sadly to date there has been little response to the task group's findings.

The second example, Khayelitsha, is a peripheral urban township of approximately one million inhabitants. KERIC (Khayelitsha Educational Resource Information Center) has started a project to empower local people through informing them of more sustainable life-styles including energy efficient housing. This will target school youth and the surrounding community, and will educate people and provide them with hands-on experience which demonstrates principles of sustainable development.

In 1994, the National Parliamentary White Paper on Housing emphasized the need to improve the thermal performance of housing, for example, the provision of basic insulation, including the installation of ceilings which will have real and quantifiable benefits in Cape Town. The Energy and Development Research Center has been involved in projects which examine the energy efficiency of houses in Cape Town. The Center is working with the Marconi Beam Development Trust informal settlement, where it is in the process of building more energy efficient houses in the Marconi Beam Affordable Housing Project's housing upgrade program.

Earthlife Africa is a non-profit voluntary organization, also affiliated to the CSC, which focuses on environmental activities. Earthlife Africa believes in clean production which uses the minimum amount of energy and raw materials to make the product. They have been successfully lobbying for the introduction of lead free petrol and the rationalization of the urban transport system.

Atmospheric pollution

Cape Town relies on its powerful "South Easter" winds, referred to as the "Cape Doctor", to blow out to sea the more obvious evidences of industrial and transport pollution as well as the use of wood fuel for low-income households' thermal requirements. However, in the winter, when the "Doctor" is not around, cold clear days result in inversions that leave a thick brown haze over the city and surrounding areas. The Caltex refinery is thought to have been a major contributor to this phenomenon.

The Caltex refinery in the Northern part of the city has for years been charged with unacceptably high levels of pollution. CBOs, along with environmental NGOs, Chemical Workers' and Industrial Union, have begun a campaign to make Caltex a more responsible neighbor and employer.

Caltex is situated in an area that is not far from residential areas. The residents complain of asthma and other chest ailments, and they object to the thick layers of black soot that cling to every exposed surface. For years, Caltex has claimed that there has been no evidence of its being responsible for these complaints. Unfortunately, the residents did not have sophisticated laboratories and professional chemical engineers to prove their claim, and they insisted that Caltex should be the one to prove that they do not cause those problems.

The issue of potential accidents has also been a focus of attention. There has been little public information about toxic releases from Caltex and other such plants in South Africa. The apartheid government's complicity with key-point or strategic industries provided a shield behind which Caltex and others could hide, despite real health dangers posed by the refinery's activities. Legislation protected Caltex's secrecy. With the change in government, Caltex realized it would have to engage public participation in its effort to get a new permit for its refinery's expansion, and it convened a review panel. Caltex did not invite the Union to join the panel, and made it clear that the panel would have no real power as its role would be one of public approval: it would make suggestions, but Caltex would be free to ignore these. Such a set-up was to be considered as adequate to meet the public participation requirements of the permit.

Most CBOs, NGOs (including the CSC) and the Union quickly objected to these arrangements, forming a unified front known as the Anti-Pollution Alliance that called Caltex to provide all information about the refinery impacts and to commit itself to negotiate a binding agreement, prior to the constitution of a body of civil society representatives that would engage Caltex on the operation of the entire refinery and the use of its products. For three months Caltex remained silent, refusing to volunteer information and offering no response to those demands. Meanwhile, the community began to organize regular meetings from which a powerful and diverse coalition and carefully considered strategy has emerged.

Although Caltex agreed that it would negotiate a contract with the community, it did not hand over information about the refinery, particularly the baseline of emissions. It claimed that much of the information might prove to the embarrassing so it would be only released to an appropriate body, and not to the general public.

Community organizations

The community also began a process of educating themselves about the refinery and the laws governing it. The coalition grew and now has the support from the Legal Resource Center and a consultant to the coalition, a chemical engineer from the Peninsula Technikon. The community became familiar with chemicals like sulfur dioxide, and learned more about Caltex's need to have public participation in order to get its permit. This gave the community power and the ability to insist that it would only participate in a process that was meaningful.

In order to obtain emissions data, the community resorted to the government. The Department of Water Affairs and Forestry (DWAF) agreed to provide all information about water pollution from the refinery. The Legal Resource Center undertook to get the relevant airborne emissions data, but the Department of Environmental Affairs and Tourism did not immediately comply. With the Deputy Minister's intervention, Caltex was compelled to be committed to openness and community relations. Caltex replied to the Deputy Minister's request, thereby averting a legal showdown. Caltex met with Legal Resources Center and offered for the first time to supply information. This was the first victory. There is an intention to convene discussions with the government regarding the nationwide regulation of all pollution from all refineries.

Finally, Caltex responded, committing itself to a binding agreement to clean up their activities, initially reducing emissions to 80 per cent. This is at the core of a first phase agreement, in which emissions standards are linked to the parent company's (Texaco) USA emissions standards. A proposed second phase relates to regular evaluations of the Caltex operation and new performance criteria compared to "best-available technology" being negotiated. This outcome has been a second victory and an important precedent in community-industry relations throughout South Africa. Companies always seem to prefer to resist rather than permit themselves to engage in contacts with interested and affected parties, something which they have been allowed to get away with for a long time.

Biodiversity and the metropolitan open space systems

The implementation of a Metropolitan Spatial Development Framework is being coordinated by the Metropolitan Planning Authority. The Cape Metropolitan Council has sought the input of a range of stakeholders, including NGOs and CBOs, in the development of metropolitan environmental management. One of the main objectives of the exercise is to guide the formation of a forum to advise on the physical development of the Cape Metropolitan Area, including the Metropolitan Open Space System (MOSS).

The creation of a planned open space system within the urban area is essential in order to protect existing open spaces of biological value, to provide new open space opportunities where necessary, as well as a more effective system of evaluating the use of land within the metropolitan area. An effective MOSS should have a number of beneficial results which will contribute towards a more sustainable city by means of:

1 Economic development and social mobility, in particular, by providing better access to outdoor recreation and safe public space for the entire urban population, including the poor.
2 Protecting and enhancing tourism resources.
3 Providing opportunities for environmental education.
4 Providing open space areas which may assist in job creation.
5 Promoting urban agriculture, tourism, informal trading and public works programs.
6 Safeguarding ecological resources and processes by designing an open space system which promotes better stormwater management, air pollution control, solid waste management, increased biodiversity and habitat protection.
7 Complementing a more sustainable urban structure.

It has been argued in the MSDF that urban areas should be more compact and should contain higher density development as well as focusing on a public transport system in a more consolidated and continuous pattern of open space which allows for multiple uses to occur, which is a more appropriate development strategy for the current sprawl.

In addition to participating in a metropolitan wide planning initiative such as the MSDF, the Central Substructure has had a long history of involvement in open space and recreation planning projects at municipal and more local levels. Unfortunately, the MSDF does not yet enjoy political support, and is often overlooked in favor of business-as-usual *ad hoc* developments.

The Central Substructure is actively seeking to establish partnerships with NGOs and CBOs in line with the RDP, to make these initiatives successful by means of incorporating local knowledge and opinion by: identifying local issues, needs and values; providing specialist advice; identifying valuable habitat and rare plant species; providing public support for protecting and enhancing open spaces; establishing links between the local authority and the broader public; information exchanging and networking; assistance in securing private funding for open space projects; and participation in implementing and monitoring projects.

The Central Structure developments have included those in Liesbeek River walkway, Kenilworth Racecourse, Wolfgat Nature Reserve, Rondebosh Common, and Newlands Forest. The CMA, incorporating Cape Town, Strand, Stellenbosh, Atlantis and Paarl, supports at least 294 plant species which are in immediate danger of extinction, as well as numerous amphibian, insect and other Red Data Book species. The numbers of Red Data Book species of butterflies, amphibians and reptiles are the highest in the Southern part of South Africa centered on the Cape Peninsula and adjacent lowland.

The CMA is probably one of the fastest developing areas in the country and is expected to accelerate in response to Cape Town's bid to host the Olympics. In addition, the extra land and space requirements inherent in such a far reaching proposal could exacerbate land allocations pushing formal low-income settlements further out of the city.

Waste management

Within most South African towns and cities there exist vastly disparate standards of waste collection. Higher- and middle-income areas enjoy frequent waste collection and street cleaning. In low-income areas, regular domestic waste collection may be infrequent, or non-existent. Without adequate waste collection services, residents are faced with little alternative but to dump waste and litter on street corners or empty lots. To compound matters, it is these communities who often bear the brunt of the problems caused by others who dump their waste illegally because of ignorance, disrespect or avoided high landfill charges.

As a society of consumers, the Cape Metropolitan Area generates vast quantities of waste, which is estimated currently as 650,000 tons of domestic and commercial waste per annum. The amount of waste is about to increase if socio-economic development is accompanied by increases in the consumption and careless discarding of waste. Industrial waste has an even bigger environmental impact. The CMA generates an estimated 97,400 tons per annum of industrial processing waste, which excludes an estimated 12.5 million tons per annum of mining and power generation waste.

If mismanaged, all waste is potentially hazardous. Similarly hazardous is industrial development without concurrent efforts at waste reduction. There is very little data on waste quantities, long-term behavior of landfills and the waste management requirements of a rapidly growing household and industrial sector, making an accurate analysis difficult. The region has a shortage of safe properly managed landfill sites. Some sites are inappropriately located close to water courses or settlements; in fact, some settlements have sprung up on reclaimed landfills. At best, these sites are unpleasant for nearby residents, and at worst they give rise to potentially serious health and safety risks.

An important aspect of waste management missing from current management is the lack of encouragement for households to separate and reduce waste streams. For example, in Cape Town a flat rate of 23 per cent of land tax (property rates) is utilized for solid and liquid waste management no matter how much refuse is discarded.

That is the reason why waste management practices should be based on minimum requirements articulated in the Department of Water Affairs and Forestry-DWAF's principles of integrated waste management strategies for the Western Cape. Landfill problems include medical and abattoir wastes; scavenging which exposes people to health and safety risks; contaminated liquid leachate escaping from the site may pollute surface and ground water; and windblown surface materials causing unsightly litter. The air around landfill sites is polluted by smoke from burning waste and evaporating fumes from ponds containing hazardous toxic waste. Few future landfill sites have been identified.

As a result of existing pollution and other environmental problems as well as of more stringent legislation recently introduced, several landfill sites have been closed and others have been given closure orders by the DWAF. A number of the currently operating sites have a lifespan limited to less than five years.

The Coalition for Sustainable Cities is aware of the past policies the local authorities have been relying on to approach waste management, which concentrates on waste collection, transfer and disposal. They drafted some policy principles for a way forward which include:

1 At national level, the government should establish one integrated waste management policy.

2 All waste management should be regulated by a single act of parliament.
3 Local and regional authorities should be encouraged/forced to implement a waste management approach to managing the waste stream in their areas of jurisdiction. This would include setting up schemes to assess and encourage better waste management.
4 Reduction at source, recycling, reuse, and waste treatment.
5 Local authorities should investigate the use of financial incentives to encourage waste minimization.
6 Local authorities should be encouraged to launch educational programs to make people aware of the extent of the problem areas.

The alternatives which are available to local authorities should involve communities in developing an integrated waste management policy for the region.

Water

South Africa is a dry country. Water is a limited resource nationally, and this also applies to Cape Town, the water supply of which will meet the current growth in demand for only a few more years. The CSC believes that every person within the CMA has the right to clean and safe water, which should be readily available and equitably distributed to all. In order to meet this demand for water within the city, while protecting the natural systems that supply it and the freshwater resources of the region as a whole, the CMC needs to manage water on a catchment basis. Nine-eight per cent of the Cape Metropolitan Area's water is transferred from rivers that are outside the boundaries of the CMA such as Steenbras, Riviersonderend, Eerste Berg and soon the Palmiet, the last undammed catchment, will also be a supplier.

Local engineers and the Department of Water Affairs and Forestry are looking ahead to the future of water supply to the CMA. In 1987, a systematic study of the water supply options for the CMA was undertaken, which was known as the Western Cape System Analysis. The results of this study are currently being assessed by the communities and the different supply options are being evaluated in order to rank them.

Most of the water supply options for the CMA involve rivers of great natural and cultural value. The quality of water in upper catchments is unspoiled. They are ideal sites for dams, and evaporation is also lower higher up the catchment. There is a great deal of concern that the

Western Cape will gradually lose all its pristine upper catchment as the rivers are dammed.

There is an unequal distribution and implementation of water supply and sanitation works in the CMA. This is leading to the irreversible contamination of underground water resources on the Cape flats.

There is also a pollution problem around the coastline of the CMA. There is not sufficient control over water that is allowed to flow into the sea from sewage works and stormwater runoffs, including industrial effluent. More urgent water quality control measures need to be determined and enforced.

The Coalition for Sustainable Cities has drafted policy options on water management in the CMA. They include:

1   Demand management should be favored above supply management. This would be facilitated through increased public participation in water management policies.
2   Holistic catchment management would include clearing of alien invasive in order to increase runoff.
3   The city should decrease leakage and water wastage because 23 per cent of water consumed within the municipal area is unaccounted for.
4   Pricing of water should be based on a sliding scale which takes quantities consumed into account, favoring those who use less.
5   Water is a precious resource and higher prices could deter wastage particularly in agriculture and industry.
6   People need to be educated on the functioning of fresh water systems and the sources of the city's fresh water.
7   Water managers need to recognize the variability supply from resources, and manage water accordingly.

Conclusions

There is an enormous scope for the improvement of environmental management policy in the Cape Metropolitan Area. Governance is currently in flux and the powers and functions of local government are in the process of being defined. This process has lent itself to the development of partnerships between government and civil society. Politicians are accessible. Opportunities to develop partnerships are being explored and realized. But much of the work in the CMA and in the Western Cape region depends upon the commitment of the National Party (which is in the majority) to participatory democracy. The watchdog role of the NGOs and CBOs in this regard, and others related to abuse of political power, needs strengthening in South Africa's new democracy.

The Coalition for Sustainable Cities is driving a proactive bid to establish policies and institutional structures through which civil society has vehicles to advise the new democratic local government. Policies that affect environmental justice issues are in the process of being developed through a series of workshops which will draw out communities' problems, formulation of and confidence in possible solutions.

Fundamental to the aims of a more sustainable development in our city are: improvement in access to adequate and affordable services; tariffs that reflect the real costs of natural resources and which encourage the efficient use of these resources and penalties for over consumption; procurement of goods and services which demand scrutiny of environmental performance; land use planning that encourages urban densification and associated improvements of public transport systems; and water and energy management on a least-cost basis.

There is some urgency in addressing inequitable access to basic needs, such as adequate housing and service provision. Similarly, a major lever is the recognition that the long term future of the Cape Metropolitan Area will be developed around its attraction as a destination for tourists interested in the natural beauty that Cape Town has to offer.

One way currently being utilized to develop policy is through partnerships that afford an opportunity to civil society and local government to work together in solving local environmental problems. Partnership as a strategy may provide a real chance for local ownership of a transparent policy process, a more aware and environmentally interested public, and the resourcing of first-lines of environmental defense. The CSC, though under-resourced, is playing a key role in the creation of just such an agenda, learning from pilot projects, international experiences and case studies such as the Model Communities Program.

Note

* This paper was compiled from contributions from the NGO, CBO and local government sectors in Cape Town by Vuyiswa Tindleni, of Langa Civic, which is affiliated to the Coalition for Sustainable Cities (CSC), Cape Town, South Africa.

4 Environmental problems in cities in the South: sharing my confusions

David Satterthwaite

Introduction

Much of the literature on environmental problems in cities in the South has a Northern bias in what it considers to be environmental problems and priorities for environmental action. It also generalizes about environmental problems, drawing from the small and representative sample of cities about which there is detailed and easily available environmental data. And it often includes statements or assumptions about the links between environmental problems and the size, density and population growth rate of cities, that are only partially true or incorrect.

This paper has three main aims. The first is to ask what the main environmental problems in cities in the South are and to suggest that this discussion has to specify not only "whose problems", but also which kind of environmental problems. The second is to inquire why so little attention is given to who benefits from cities' environmental problems. The third is to discuss what links exist between the scale of environmental problems in a city and the city's size, density and the rate at which it is growing.

The main environmental problems

Environmental problems in cities in Latin America and Africa tend to be seen through eyes that have Northern perceptions and preoccupations. Much of the research and published work on environmental problems in cities in Africa and Latin America (and Asia, for that matter) is written by specialists from Europe or North America, or by specialists from the South who were trained in Europe and North America. Unsurprisingly, the environmental problems they see, measure and write about are the environmental problems that they were trained to see, measure and write about.

Many do not see the environmental health problems that are often the main causes of premature death, disease and injury in the city as true environmental problems. As in the North, they concentrate on issues related to resource use and waste or to ambient air pollution. They tend to publish papers which give more attention to the loss of agricultural land to urban expansion or to cities' ecological footprints as the major environmental problems - while the environmental hazards that are the main causes or contributors to premature death, illness and injury are hardly mentioned.

In discussing environmental problems in cities, it is important to distinguish between at least to the following five different categories:

1 Environmental hazards within the human environment - including biological pathogens (or their vectors), chemical pollutants (in the home, at work, in the wider city), physical hazards and psycho-social stressors.
2 High levels of use of those renewable resources that are only renewable within finite limits. For finance, agricultural and forest products are only renewable resources if the eco-systems in which they are grown are not degraded. Fresh water resources are also finite; in the case of aquifers, human use often exceeds their natural rate of recharge and such levels of use are unsustainable.
3 High levels of use for non-renewable resources; for instance, fossil fuels, metals and other mineral resources. Most of these resources (especially the fossil fuels burnt for heat and power) are consumed when used, so finite stocks are depleted with use. Others are not "consumed" since the resource remains in the waste - for instance, metals used in capital and consumer goods. But for most non-renewable resources, there are energy and cost constraints to recovering a high proportion of the total amount used from waste streams. These may be considered as partially "renewable", with the

extent defined by the proportion of materials in discarded goods which can be reclaimed and recycled.

4   High levels of generation of non-biodegradable wastes which have serious ecological implications - for instance, greenhouse gases and the large volumes of non-biodegradable toxic hazardous wastes that have to be stored and kept entirely isolated from eco-systems because of the damage to eco-systems and to human health they pose.

5   Over-use of the renewable sink capacity, the finite capacity of eco-systems to break down biodegradable wastes. Although most wastes arising from production and consumption are biodegradable, each eco-zone or water body has a finite capacity to break down such wastes without itself being degraded.

Too little attention is generally given to the first of these categories - and within this category, too little attention is given to biological pathogens. Yet it is the pathogens in the air, water, soil or food that cause infectious and parasitic diseases that are much the most serious environmental problems in most cities in the South. As a recent WHO document stated:

Today, more than a third of the urban population in Africa, Asia and Latin America live in housing of such poor quality with such inadequate provision for water, sanitation, drainage, garbage collection and care that their lives and their health are constantly under threat. In such circumstances, it is common for one child in three to die before the age of five and for virtually all infants, children and adults who survive to have disease burdens many times higher than they should. For instance, the health burden per person from diarrhoeal diseases caught in 1990 was around 200 times higher in sub-Saharan Africa than in the North, and less overcrowded housing with adequate provision for water, sanitation and the safe preparation and storage of food would enormously decrease this health burdens. Disease burdens from tuberculosis and acute respiratory infections (each along with diarrhoeal diseases, among the largest causes of death worldwide) are generally much increased by overcrowding. Many accidental injuries happen when there are three or more persons living in each small room in shelters made of flammable materials and there is little chance of providing occupants (especially children) with protection from open fires or stoves.[1]

One possible reason why many environmental specialists miss these environmental problems is that these are no longer serious environmental problems in cities in the North. The almost universal provision for piped water, sanitation and drainage and much improved health care provision in the North has so reduced the contribution of biological pathogens to serious illness or premature death that these are no longer seen as environmental problems.[2] Yet only 100 years ago, when the deficiencies in provision for piped water, sanitation, drainage and health care were similar to those in many cities in the South today, even prosperous European cities had infant mortality rates that exceeded 100 per 1,000 live births; in Vienna, Berlin, Leipzig, Naples, St. Petersburg and many of the large industrial towns in England, the figure exceeded 200 and in Moscow, it exceeded 300[3] - while infectious diseases were the main cause of such high infant mortality rates.[4]

In other areas too, the literature on environmental problems in cities in the South seems too influenced by Northern priorities. For instance, when considering the impact of chemical hazards on city populations, there is far more literature on ambient air pollution that on indoor air pollution and on occupational exposure - although in many cities indoor air pollution and occupational exposure are likely to have more serious health impacts.

Whose environmental problems?

Much of the general literature on cities' environmental programs also fails to ask who is most affected by these problems. However, most studies on infectious and parasitic diseases and morbidity and mortality show that these are concentrated among the low income groups - be they children, adults in crowded, unhygienic conditions or workers in particular occupations.[5]

Low-income groups are the least able to afford the homes that protect against environmental hazards, that is, good quality housing in neighborhoods with piped water and adequate provision for sanitation, garbage collection, paved roads and drains. In addition, higher-income groups will generally have less dangerous jobs and work in occupations where occupational hazards are minimized. If there was sufficient information available to construct a map of a city, showing the level of risk from biological pathogens, chemical pollutants and physical hazards, the areas with the highest risks would generally coincide with the areas with a predominance of low-income groups.

Thus, virtually all environmental health problems in urban areas have a social, economic or political underpinning in that it is social, economic or political factors which influence or determine who is most at risk, why they are risk, why nothing has been done to lessen the risk the risk and why they cannot obtain the treatment and support they need, when illness or injury occurs.[6] To give but one example, the high incidence of diseases associated with contaminated water in most poor urban communities is an environmental problem in that the disease-causing agents infect humans through the water they ingest, but this high incidence can also be judged to be a political problem since nearly all governments and aid agencies have the capacity to greatly reduce current levels of morbidity and mortality by improved provision of water, sanitation and drainage and for health care (to allow quick and effective treatment, when they fall ill).

It can also be judged to be a social or economic problem in that it is often lower-income groups' limited means to pay for accommodation that underlies the fact that they live in housing that lacks adequate provision for water, sanitation and drainage. If only a proportion of the housing stock in any city is adequately served with piped water and provision for sanitation and drainage, inevitably it is market forces that are the major influence on who has access to such housing. This makes it difficult to isolate the impact of environmental factors on health as distinct from other factors.

There are also groups within "the poor" and wider city population there are particularly vulnerable to certain environmental hazards, and another level of analysis is needed to consider who within poor groups (and wider population) suffers most from environmental problems and how their vulnerability can be reduced or removed. For instance, there are particular occupational groups with high levels of exposure to environmental hazards such as toxic chemicals, dust and noise.

There are also those groups within the population who are particularly vulnerable to an environmental hazard because they are less able to avoid a hazard or are more affected by it. For instance, those with weak body defenses are particularly vulnerable to many biological pathogens, with the strength of a body's defense being much influenced by age and nutritional status - and by acquired immunity against diseases (including those diseases for which protection can be achieved from vaccination). Infants and young children are at greater risk than other children and adults because of immature immune systems.

Those suffering from under-nutrition also have less effective immune systems and are less able to recover quickly from diseases or injuries. For exposure to chemicals, the health impact is influenced by age, activity when exposed and health status at the time of exposure - and again infants and young children are particularly vulnerable to many chemicals.[7] Pregnant mothers and their foetuses are also particularly vulnerable to many biological pathogens and chemical pollutants.[8] Those with limited mobility, strength and balance are particularly vulnerable to many physical hazards, including many elderly people and infants and young children. There are certain tasks which expose those undertaking them to higher levels of risk. For instance, the person within a household who undertakes most of the cooking and domestic tasks (usually women and girls) are also more at risk from indoor air pollution.[9] Thus, in households where cooking is done on open fires or inefficient stoves using coal, wood or biomass fuels, and with poor ventilation, those who cook and spend most time in the house have much greater exposure to the high levels of indoor air pollution.

As well as considering who within the city population is most at risk from environmental hazards, consideration must also be given to the health of those living outside the city or as yet unborn, since this can be threatened by the demand of resources or the generation of wastes by city-based enterprises and consumers.

We have also to consider the environmental problems that face those who live outside the city, but who are affected by resource demands or wastes arising from city-based production or consumption. For instance, the problems faced by those whose livelihood comes from fishing and whose fisheries are being damaged or destroyed by water pollution from the city. Or the farmers and rural communities whose local water resources are being appropriated by city or regional authorities to supply the city's inhabitants and enterprises.

These are also the people living far away from cities whose health is damaged by environmental hazards that relate to the work they do producing goods for the city. This problem has international dimensions. For instance, one environmental problem to which consumers living in wealthy cities in the North inadvertently contribute is the poisoning of farm workers in rural areas in the South who suffer from pesticide poisoning producing crops that are sold in the North. Many of the environmental impacts created by the demands for goods by consumers and businesses in wealthy cities are transferred to distant eco-systems, with the environmental health costs being borne by inhabitants of these distant eco-systems.

One reason why London, for example, can maintain a high quality environment around its built up area is that many of the goods that its population uses that require heavy inputs of water, energy and dangerous chemicals are imported from abroad. This is a natural consequence of changes such as tighter environmental and occupational health regulations in the North and the moves in the North to protect natural landscapes around cities, so they are no longer available for food production. It is facilitated by an increasingly globalized production structure and the relatively low cost of oil-based fuels - and, of course, by cheap labor and lax or inadequate environmental and occupational health controls in the producing countries.

Finally, there are the environmental programs that current city-dwellers are passing onto future generations - for instance, in the health and ecological costs that global warming is likely to cause for future generations or the loss to future generations from the declining biodiversity. Thus, a consideration of environmental programs in cities in the South has not only to consider who within the city population is most affected, but also who outside the city is affected. It must include a concern for those as yet unborn. It must also be recognized that as cities become more wealthy, so too does the ability of their inhabitants and businesses to transfer environmental costs for their own homes, neighborhoods and cities to other people and other places.[10]

Who benefits from environmental problems

Although there is a general literature about who benefits from environmental problems, there is not much detailed information about this subject in cities of the South. Yet it is clear that certain powerful vested interests are the main beneficiaries of lax or no government action on the environment. Some points are outlined here that deserve a more detailed consideration.

In regard to toxic wastes, the owners and operators of heavy industry and other industries producing hazardous wastes benefit from no regulation governing their storage and disposal. The safe handling and disposal of toxic wastes is very expensive, so there are large incentives to ignore any regulatory system or to cheat. The costs of destroying or safely disposing of hazardous wastes may be US$ 1,000 or more a ton. Dumping them on some landfill site costs very little. Dumping them into a nearby river or lake costs almost nothing.

In regard to industrial pollution, it is clear that the major industrial concerns benefit from three things:

1 The cost savings from not having to install pollution control equipment or replace old equipment because of lack of pollution control or its enforcement.
2 The cost savings arising from an absence of regulations on occupational health and safety or a lack of their enforcement.
3 In many cities, plentiful and usually cheap supplies of pipes water and other forms of public infrastructure and services (roads, drains,

electricity, telecommunications, etc.) for which they may pay a prize that is well below the real cost of providing them.

In effect, the owners of industrial concerns appropriate various environmental resources which are generally considered as public resources or open access resources: clean air (into which they dump air pollution), freshwater (generally priced so low that they have little incentive to use if efficiently) and local sinks (where they dump their untreated wastes). There are also groups who make money out of the fact that there is inadequate or no public provision for water supply, garbage collection and primary health care, as they provide commercial services for these.

In most cities, there is also likely to be a considerably body of landowners who benefit from the inadequacies of land use control, as they can develop and sell land for uses that are inappropriate or with inadequate provision for infrastructure and services. In any fast growing city with a prosperous economy, there is so much money to be made from acquiring and developing land to meet the demands of businesses or upper-income groups that it is very difficult to ensure that the public good is protected and the needs of low-income groups taken into account.

Middle- and upper-income groups and business generally benefit from the environmental problems outlined above because they are not required to contribute to the cost of remedying them, the lack of government action to address them means that they need pay little or no local taxes. As long as the environmental problems mostly affect the homes, neighborhoods and work places of low-income groups, and not the homes of middle- and upper-income groups, they need not be concerned about them. Although the possibility of disease epidemics in cities which also affect middle- and upper-income groups has been raised as one means to persuade wealthier groups to contribute more to environmental improvement - as was the case in 19th century Europe and North America - in reality, wealthier groups have the means to protect themselves.

In addition, middle- and upper-income groups generally have the political influence and connections to ensure that whatever public funding is directed to environmental improvements is used primarily for their benefit. Middle- and upper-income neighborhoods are far more likely to receive public provision of water, sewers, paved roads, drains and garbage collection than poor neighborhoods. Middle- and upper-income groups are generally the main beneficiaries of government subsidies for housing or for housing finance. As a result, public investment and action (and inaction) may reinforce the disparities in the environmental risks faced by high and low income groups.

In many cities, middle- and upper-income groups have also appropriated for themselves access to open space and natural beauty. For instance, in many major cities, country clubs and golf courses are the only easily accessible unbuilt up areas and access to these is restricted to members. In coastal cities, many of the best beaches nearby may have been privatized - usually illegally - or incorporated into clubs with access restricted to members.

Of course, this section risks making the kinds of generalizations that earlier sections criticized. There are many cities in which the lack of public provision for infrastructure and services imposes high costs on private businesses as they have to invest in their own water supply system, electricity generators and telecommunications systems.[11] There is also the fact that the increasing competition between cities to attract foreign investment can help to mobilize more action to address cities' environmental problems. But it might help to mobilize more action to address cities' environmental problems if there was a greater understanding of who benefits from these problems not being addressed.

The links between environmental problems and the size, density and growth rate of cities

Links between city size and environmental problems

There is certainly no obvious link between the size of a city and the extent of the first of the environmental problems listed above, that is, environmental hazards within the human environment. Since much of the illness, injury and premature death arising from environmental hazards in cities in the South can be linked to inadequate provision for water, sanitation, drainage, garbage collection and health care, it is in the cities where provision for these is most inadequate that environmental health problems are most serious.

Although there is too little detailed data on the proportion of each city's inhabitants with adequate provision for each of these, in general it seems that a higher proportion of the population is adequately served in the largest cities.[12] This would not be surprising given that the largest cities are usually more prosperous and more politically important (especially capital cities). And when judged for the whole of the South, of course, there is the high concentration of the world's largest cities in the world's largest economies, as indicated in Table 1.

70

Table 1

The association between the world's largest economies and largest cities, 1990

	10 million plus cities	5 million plus cities	"million cities"
Worldwide	12	33	381
In the world's five largest economies	6	11	103
In the next ten largest economies	5	11	67
In the next ten largest economies	0	4	77
In the rest of the world	1	7	77

Source: Satterthwaite, D., *The Scale and Nature of Urban Change in the South*, London, IIED (mimeo), 1996.[13]

The cities where environmental health problems are likely to be the most serious and to affect the highest proportion of the population are the smaller and less prosperous cities in lower-income countries or in the lower-income regions of middle-income countries. Most will have no sewer system at all, while few will have a well managed piped water system that reaches most of their inhabitants.

There are environmental problems that generally increase with city size - for instance, from industrial emissions, although at some point this may decline in importance among the larger cities either as more effective emission controls are imposed or as the large cities lose their heavy industry and/or polluting industrial base. There are also the environmental problems that almost always increase with city size - for instance, pollution linked to motor vehicle emissions, although the extent of the problem in any large city will be considerably influenced by such factors as local meteorological conditions, the configuration of the city, the quality of traffic management, quality of the motor vehicle fleet, and the extent to which people can move into and around the city without using a private automobiles. Certain cities in the South have serious problems with air pollution related to motor vehicles (Mexico City, Teheran, Santiago, etc.), but these are often related not only to local emissions but to local meteorological and topographic features.

There is a set of environmental problems that almost always relate to the size of the city - and these are the other four categories of environmental problems listed above, which collectively define the "ecological footprint" that the city's producers and consumers create in terms of their demand for goods and for the waste sink capacities of eco-systems and the wider ecosphere. The fact that the world's largest cities

71

have among the largest "ecological footprints" is also a reflection of the fact that they are heavily concentrated in the world's largest economies (see Table 1) and that they concentrate a high proportion of the world's highest consumers and a considerable proportion of the world's non-agricultural production.

Links between cities' growth rates and environmental problems

Again, the link between the speed with which a city has grown over the last few decades and the extent of the environmental hazards for its population is likely to be tenuous, in that many of the fastest growing cities are also the cities with the fastest growing economies and the cities where most investments have been made in improving provision for piped water, sanitation, drainage, garbage collection and health care. For instance, in Brazil, many of the most rapidly growing cities of the last two or three decades are also those with among the best provision for water and sanitation and with life expectancies that are much higher than cities that have grown much more slowly.[14] Since Curitiba in Brazil is so often held up as a city which has been much more successful than most other cities at achieving high environmental standards, it is worth noting that this is also one of the most rapidly growing large cities in the world over the last few decades. São Paulo and Mexico City may be held up as "horror stories" in regard to their urban environment, but in fact both cities have a well-above average performance in provision for safe and sufficient water supplies and provision for sanitation compared to all cities in the South.

It is also worth noting that several of the world's fastest growing cities in this century are in the United States. Cities such as Dallas, Las Vegas, Phoenix and Miami figure among the world's most rapidly growing cities during the 20th century - and these are not cities characterized by most of the populations lacking access to safe and sufficient water supplies and provision for sanitation. Clearly, rapid population growth in itself is not an useful explanation for the scale of the environmental hazards to which the city's population is exposed.

Environmental problems and density

A senior World Bank environmental specialist claimed recently that "the very density and form of urban areas are contributing factors to, if not determinants of, a city's environmental quality."[15] But this does not accord with the fact that some of the most desirable, expensive urban residential areas in the world combine very high quality environment with high densities, and the fact that many illegal or informal settlements with very poor quality living environments do not have a particularly high density. It also misses the point that high densities can bring substantial environmental advantages.

For a residential area, its population density does not necessarily give much idea of its environmental quality (including the level of risk from environmental hazards). For instance, a density of 1,000 persons per hectare living in one-story shacks without piped water and provision for sanitation and drainage will constitute a very poor quality and dangerous environment. Yet 1,000 persons per hectare could live in a high quality living environment - for instance, in apartments in very high quality housing such as in one of the most expensive five-story Georgian terraces in Chelsea, in London.

Cities' population density also do not give much idea of environmental quality since, again, the environmental quality is too dependent on the quality of the housing stock and of provision for maintaining the streets and public spaces and for removing household and human wastes. Some of the most pleasant and valued parts of London, Copenhagen, Paris and Amsterdam (among many cities in Europe) have population densities as high as many of the most dangerous and polluted urban environments in the South. The critical link between environmental quality and density in urban areas is that, in general, the higher the population density, the more a good environmental quality is dependent on good public or community provision of infrastructure and services and good environmental management.

Cities with a relatively high density also bring many environmental advantages over low density including:

1   Much lower costs per household and per enterprise for the provision of piped, treated water supplies, the collection and disposal of household and human wastes, and most forms of health care and education. It also makes much cheaper the provision of emergency services - for instance, firefighting and the emergency response to acute illness or injury that can greatly reduce the health burden for people affected. The concentration of people and production may present problems for waste collection and disposal, but these are not problems that are insuperable, especially where a priority is giving to minimizing wastes.

2   A greater range and possibility for efficient use of resources, through the reclamation of materials from waste streams and its reuse or recycling, and for the specialist enterprises that ensure this can happen safely. The collection of recyclable or reusable wastes from homes

and business is generally cheaper, per person served, in high density settlements.

3 A reduced demand for land relative to population. In most countries, urban area take up less than one per cent of the national territory.

4 A much greater potential for limiting the use of motor vehicles, including greatly reducing the fossil fuels they need and the air pollution and high levels of resource consumption that their use implies. High density cities allow many more trips to be made through walking or bicycling and they also make possible a much greater use of public transport and make economically feasible a high quality service.

These caution against assumptions that high densities are necessarily associated with environmental problems.[16]

Changing nature of environmental problems as cities get larger and/or wealthy and/or more competently managed

One of the key insights about how environmental problems in cities change in relation to their income is the comment by Gordon McGrannahan and Jacob Songsore that, as cities get larger and the residents wealthier, so do they can transfer their environmental problems away from their homes and neighborhoods to other people and regions. They also transfer responsibility for these problems to someone else:

> ... many environmental services such as piped water, sewerage connections, electricity and door to door garbage collection not only export pollution (from the household to the city), but also shift both the intellectual and practical burdens of environmental management from the household to the government or utility.[17]

The scale and relative importance of the different environmental problems listed above change if a city's (and country's) economy become more prosperous and if more investment can be made in infrastructure and services.[18] In general, the higher a country's per capita income, the higher the proportion of urban (and rural) population with safe water and adequate provision for sanitation, and the larger the municipal wastes - but also the more comprehensive the service to regularly collect such wastes. If, in addition, provision for health care also improves as a city becomes more prosperous, then it is likely that environmental health problems will reduce rapidly.

However, certain other environmental problems also increase with per capita income. These include two that increase environmental hazards within the human environment:

1 Concentrations of particulate matter in the air in urban areas - although in many wealthy cities this is a problem that has peaked and is now in decline, as emissions have been cut.
2 Concentrations of sulfur dioxide in the air in urban areas - although in most wealthy cities there have been rapid decreases linked to such changes as reduced use of high sulfur coal in industries and domestic heating.[19]

In many cities, problems of air pollution linked to motor vehicle exhausts are also increasing.

Although there are general tendencies that have been shown in international comparisons, there are also large variations in many such indicators between cites in countries with comparable per capita incomes. In general, the percentage of the population with water piped to their plot increases in cities, the higher the per capita income of the country in which they are located. However, there are also large variations between cities in nations with comparable per capita incomes, at least in countries with per capita incomes below US$ 10,000 in 1991.[20] There are also very large differences in per capita gasoline use among the world's wealthiest cities.[21]

This tendency for improved environmental quality to be associated with higher income was demonstrated by a study of household level environmental problems in three cities: São Paulo (the largest city in Brazil and one of the wealthiest city in the South); Accra (the largest city in Ghana, a relatively low-income country); and Jakarta (the largest city in Indonesia, a middle-income country). With surveys covering 1,000 or more households in each city, the environmental problems associated with water, sanitation, solid waste, indoor air quality and pests were found to lessen with the per capita income of the country (see Table 2). But as importantly, most household level environmental problems also tended to lessen within each of these cities, the wealthier the household.[22]

The fact that the scale and relative importance of environmental problems change the higher the per capita income of the city or nation, allows for a general qualification of urban centers according to which set of environmental problems (including environmental hazards within the human environment and the generation of biodegradable and non-biodegradable wastes, among others) is likely to be most pressing.

Table 2

Incidence of household environmental problems in Accra, Jakarta and São Paulo

	Incidence of problem (percentage of all households surveyed)		
Environmental indicator	Accra	Jakarta	São Paulo
Water			
No water source at residence	46	13	5
No drinking-water source at residence	46	33	5
Sanitation			
Toilets shared with more than 10 households	48	14-20	<3
Solid waste			
No home garbage collection	89	37	5
Waste stored indoors in open container	40	27	14
Indoor air quality			
Wood or charcoal in main cooking fuel	76	2	0
Mosquito coils used	45	28	8
Pests			
Flies observed in the kitchen	82	38	17
Rats/mice often seen in the home	61	82	25

Source: McGranahan, G. and Songsore, J., "Wealth, health and the urban household; weighing environmental burdens in Accra, Jakarta and São Paulo", *Environment*, Vol. 36, No. 6, July/August 1994, page 9.[23]

The main environmental problems vary according to the different sizes of urban centers and within nations with different levels of income, and can be classified in four principal categories: problems existing in most urban centers in most low-income nations and many middle-income nations; problems typical of more prosperous cities in low- and middle-income countries, including many that have developed as industrial centers; those problems found in prosperous major cities or metropolitan areas in middle- and upper-income countries; and problems typical of cities in upper-income countries.

In most urban centers in most low-income nations and many middle-income nations, many households are still struggling with a lack of provision for water and sanitation, but there are no problems with high levels of resource use or of wastes or of high levels of greenhouse gas emissions. In cities in upper-income countries, the main environmental problem is no longer within the city, but in the collective impact of the consumption and waste produced by city inhabitants and city businesses on regional and global resource bases and systems.

The initial environmental problems for all urban centers are those caused by the absence of infrastructure and services: the environmental health problems that arise from a concentration of people and enterprises when there is a lack of provision for piped water supplies, sanitation and drainage, solid waste collection and health care. Increased prosperity brings with it increased potential to invest in the needed infrastructure and services. For those with rising incomes, this also means rising capacity to pay for these and, if provided efficiently, most or all costs of provision can be recouped. As these initial "environmental health" problems are addressed, and as the city's economic base grows, so chemical and physical hazards grow, many of them originating in enterprises.[24]

If growing economic prosperity is based on rapid industrialization, this can mean serious industrial air and water pollution and growing problems with hazardous wastes. In cities with rapid industrial growth, there are strong pressures brought by industrial and commercial concerns to limit pollution control. Rapid industrial growth in the absence of pollution control can produce very unhealthy cities - of which Cubatão in Brazil is one of the best known examples, although environmental quality in this city has improved greatly since it was known as "the Valley of Death".[25]

As chemical and physical hazards are addressed through much improved pollution control, waste management and attention to occupational health and safety, they usually give way to environmental problems linked to high-consumption lifestyles and an increasing number of private automobiles. Problems of industrial pollution are often lessened too, through the deindustrialization of cities or more effective pollution control. The mix of air pollutants also changes, reflecting the increased role of automobile emissions and the reduced role of industry and power stations. And the large cities also have an increasing impact on their wider regions.

This movement through the four abovementioned categories can also be interpreted as an increasing capacity for individuals and households to pass the environmental costs and responsibilities for environmental management to higher levels - as noted by the quote at the beginning of this section.[26] Thus, connection to a sewer allows a household to get rid of excreta and waste waters with no hazard to the household itself and with great convenience - but the aggregation of all households' liquid wastes within a city, when disposed of untreated, can seriously damage regional water bodies. The same is true for solid wastes. If coal or wood fueled stoves are replaced with electric cookers, the air pollution burden may be transferred from inside the home and in the settlement or city to the power station.[27]

But cities do not have to successively move through those four categories. In most urban centers in relatively low-income countries, far more could be done to improve water supply, sanitation, drainage and health care using existing resources through using existing resources allocated in different ways. This also brings other important advantages - for instance, the stimulus that infrastructure improvement in residential areas brings to households investment in improving or extending housing that also creates employment.[28]

Similarly, no city has to "wait" until a certain per capita income is reached before action is taken. It is not necessary for cities to undergo a "dirty industry" phase as part of their development - as most of the older North American and European cities have undergone and as many industrial centers in the South are also undergoing. The costs of pollution control are often less important in the total costs of an industry than industrialists like to claim, especially when new industries are set up and "clean" industrial technology can be installed. In addition, a clean, well-managed city is in itself an important asset in attracting many kinds of new enterprise.

The problems of pollution and traffic congestion associated with many of the world's wealthiest cities today, most of which are "post-industrial" cities as they have very little industry, are also problems that can be avoided or much reduced with foresight. It is the way that cities expand and new settlements develop on their periphery and the spatial form of these new developments that will strongly influence the extent of private automobile use and the scale of rural land loss to urbanization.[29]

Conclusion

The extent to which urban centers in low-income nations can address the most serious life-threatening environmental problems will be much influenced by local political and economic circumstances. The uniqueness of each city must be stressed again - not least in the form of its political economy. Each city has within it a great range of actors and institutions that contribute to the city's economy and built form and that are also seeking changes in policy and approach from all agencies in government. Most are in competition with each other for scarce public resources and political favors. Many are seeking official approval for land developments that can multiply one hundredfold or more the value of land they own.

Evidence of this competition is evident in every city's local newspaper and often in national or regional newspapers, as citizen groups demand action against polluters or castigate the performance of utility companies or as those illegally settled in the city's watershed try to avoid eviction. It is the outcome of this process that largely determines what investments are made in infrastructure, which environmental problems receive priority and who benefits (and who does not). In political systems that inhibit organized action by low income groups or that are undemocratic, most investments in infrastructure and services may end up serving middle- and upper-income groups.

One possible way to greatly improve the quality and range of infrastructure and services in relatively low income countries is to draw on the knowledge and resource of low-income communities.[30] There are precedents that show how much can be achieved with modest resources.[31] But achieving this will require more resources and powers devolved to community level - and not just more responsibilities.[32] It should also be remembered that most low-income individuals and households have limited "free time" to devote to community initiatives, as most adults members have to work long hours. In addition, no assumptions should be made by external agencies as to how much community participation the inhabitants of a particular settlement want; low-income households may prefer to spend a larger proportion of their income to obtain provision for water, sanitation, garbage collection and environmental management that does not involve community management.

Notes

1 See WHO (1996).
2 However, there are some problems in the North with certain environmental health problems that are emerging or re-emerging; see WHO (1996).
3 See Bairoch (1988, p. 231) and Whol (1983).
4 Undernutrition was also a major factor.
5 See Bradley et al (1991).
6 See Hardoy et al (1992); see also Douglass (1992).
7 See WHO (1992a).

8 See WHO (1992b).

9 See WHO (1992a).

10 See McGranahan and Songsore (1994), pp. 4-11 and 40-45; see also Rees (1992).

11 See, for instance Lee (1988).

12 See Hardoy et al (1992).

13 The data on the world's "ten million plus" cities and "million cities" is drawn from *United Nations, World Urbanization Prospects: the 1994 Revision*, Population Division, New York, 1995, but adjusted when new census data that was not included in this dataset becomes available and corrected, when cities that were actually part of larger cities were included as separate cities in the original database. The data on the world's largest economies in is drawn from UNDP, *The Human Development Report*, 1993, Oxford University Press, Oxford and is based on real GDP (ppp US$); in that year, the USA, China, Japan, Germany and the Russian Federation were the five largest economies.

14 See Mueller (1995).

15 See Serageldin (1995, p. 17); see also Serageldin et al (1995).

16 See Acioly Jr. and Davidson (1996) for a discussion of density in urban development. Note should also be made that is not necessarily the concentration of population in a single city that presents the most resource efficient model - see, for instance, the discussion about the social and environmental advantages of the "compact city" versus other urban configurations as discussed in Breheny (1992) and Blowers (1993).

17 See McGranaham and Songsore (1994).

18 Work Bank (1992); Bartone et al (1994).

19 World Bank (1992).

20 See UNCHS (1996).

21 See Newman and Kenworthy (1989).

22 See Bartone et al (1994).

23 Data drawn from surveys conducted by the Stockholm Environment Institute in collaboration with a local research team in each of these cities in late 1991 and early 1992. Sample sizes were 1,000 for Accra and São Paulo and 1,055 for Jakarta.

24 Although in cities in locations where domestic heating is necessary for part or most of the year, dirty, smoky fuels used in the home can be a major contributor to ambient air pollution.

25 See Hardoy et al (1992).

26 See McGranaham and Songsore (1994).

27 Although this is only so if a thermal power station is used and it lacks emission controls.
28 See Strassman (1994).
29 See, for instance, Newman (1996).
30 See Douglass (1992).
31 See UNCHS (1996) for many examples.
32 See Douglass (1992) and McGranahan and Songsore (1994).

References

Acioly Jr., C. and Davidson, F. (1996) "Density in Urban Development", *Building Issues*, Vol. 8, 25 pages.

Bairoch, P. (1988) *Cities and Economic Development: From the Dawn of History to the Present*, Mansell: London.

Bartone, C. et al (1994) *Towards Environmental Strategies for Cities: Policy Considerations for Urban Environmental Management in Developing Countries*, UNDP/UNCHS/World Bank Urban Management Program, No. 18, World Bank: Washington, D.C.

Blowers, A. (ed) (1993) *Planning for a Sustainable Environment*, Earthscan Publications: London.

Bradley, D. et al (1991) *A Review of Environmental Health Impacts in Developing Country Cities*, Urban Management Program Discussion Paper No. 6, The World Bank, UNDP and UNCHS (Habitat): Washington, D.C.

Breheny, M.J. (ed) (1992) *Sustainable Development and Urban Form*, European Research in Regional Science 2, Pion: London.

Douglass, M. (1992) "The political economy of urban poverty and environmental management in Asia: access, empowerment and community based alternatives", *Environment and Urbanization*, Vol. 4, No 2, October, pages 9-32.

Hardoy, J.E. et al (1992) *Environmental Problems in Third World Cities*, Earthscan Publications: London.

Lee, K.S. (1988) *Infrastructure Investment and Productivity: the case of Nigerian Manufacturing - a framework for policy study*, Discussion paper, Water Supply and Urban Development Division, The World Bank: Washington, D.C.

McGranahan, G. and Songsore, J. (1994) "Wealth, health and the urban household; weighing environmental burdens in Accra, Jakarta and Sao Paulo", *Environment*, Vol. 36, No. 6, July/August.

Mueller, C.C. (1995) "Environmental problems inherent to a development style: degradation and poverty in Brazil", *Environment and Urbanization*, Vol. 7, No. 2, October, pp. 67-84.

Newman, P. (1996) "Reducing automobile dependence", *Environment and Urbanization*, Vol. 8, No. 1, April.

Newman, P. and Kenworthy, J. (1989) *Cities and Automobile Dependence: an International Sourcebook*, Gower: Aldershot.

Rees, W.E. (1992) "Ecological footprints and appropriated carrying capacity", *Environment and Urbanization*, Vol. 4, No. 2, October, pp. 121-130.

Serageldin, I. (1995) "The human face to the urban environment", in Serageldin, I. et al (eds) *The Human Face of the Urban Environment*, Proceedings of the Second Annual World Bank Conference on Environmentally Sustainable Development, World Bank: Washington, D.C.

Serageldin, I. et al (eds) (1995) *The Human Face of the Urban Environment*, Proceedings of the Second Annual World Bank Conference on Environmentally Sustainable Development, World Bank: Washington, D.C.

Strassman, W. P. (1994) "Over simplification in housing analysis, with reference to land markets and mobility", *Cities*, Vol. 11, No. 6, December, pp. 377-383.

United Nations (1995) *World Urbanization Prospects: the 1994 Revision*, UN Population Division: New York.

United Nation Center for Human Settlements-UNCHS (1996) *An Urbanizing World: Global Report on Human Settlements*, Oxford University Press: Oxford.

United Nations Development Program-UNDP (1993) *The Human Development Report*, Oxford University Press: Oxford.

Whol, A.S. (1983) *Endangered Lives: Public Health in Victorian Britain*, Methuen: London.

Work Bank (1992) *World Development Report*, Oxford University Press: Oxford.

World Health Organization-WHO (1992a) *Our Planet, Our Health*, Report of the Commission on Health and Environment, WHO: Geneva.

World Health Organization-WHO (1992b) *Reproductive Health: a Key to a Brighter Future*, WHO Special Program of Research Development and Research Training in Human Reproduction, WHO: Geneva.

World Health Organization-WHO (1996) *Creating Healthy Cities in the 21st Century*, Background Paper prepared for the Dialogue on Health in Human Settlements for Habitat II, WHO: Geneva.

5 Urban environmental management strategies and action plans in São Paulo and Kumasi

Carl R. Bartone

Introduction

The developing world's cities are currently expanding at 62 million inhabitants per year, which is equivalent to adding a country the size of the United Kingdom, Turkey or Thailand every year. Within two decades, the urban population in developing countries will double in size and, for the first time, will surpass the rural population. Fully 88 per cent of the world's total population growth will be located in the rapidly expanding urban areas - and 90 per cent of that growth will be absorbed by the developing world.

Large cities will house an increasing share of urban dwellers; one in four will live in cities of over 500,000 inhabitants, and one in ten will live in a rapidly growing number of cities of over five million residents. The largest cities, however, are slowing in growth with natural increase predominating over in-migration. In these metropolitan areas - each one comprising many municipal jurisdictions - the urban peripheries typically grow much faster than the core cities. But the highest growth rates are often observed in the secondary cities of developing countries. The fast pace of growth of intermediate cities of between 500,000 and one million population means that many such cities will be transformed into large metropolitan areas within a decade.

While reflecting on these trends, we must not lose sight of the role that urban areas play as centers of growth. Typically, cities drive the development process for a country in the early stages of modernization and give rise to more productive households and enterprises. However, pockets of poverty also spring up in cities and environmental degradation often occurs as a result of the people and their economic activity concentrated in urban centers. Perversely, in developing countries the urban poor are hit the hardest by the deteriorating physical and natural environment that surrounds them, affecting their health, productivity and quality of life. Achieving "sustainable cities" requires meeting the social and economic needs of present urban populations, especially the poor, while balancing broader environmental and energy concerns.

The urban environmental challenge

While the environmental problems facing cities are indeed dismaying, they are not insoluble. Since, however, the agenda for urban action is so broad and varied, it is necessary to choose priorities and to mobilize around a few key challenges that deserved immediate attention. In its report to the 1996 "City Summit" in Istanbul, *Livable Cities for the 21st Century*, the World Bank identified several priorities for improving the urban environment.

Extending basic urban environmental services, especially for the urban poor

In the low-income countries, overcoming the deficit in services and adequate shelter is a broad-based priority. Experience shows that effective responses are based on providing services that people want and can afford. Thus, attention must be given to understanding local demands and involving communities themselves in deciding on the type and quality of service to be provided. At the same time, supply side constraints should be removed through policy and institutional reforms and greater involvement of the private sector, including microenterprises. In addition to mobilizing community resources, successful interventions should focus on resolving land tenure issues, safety net programs, and improved municipal management which redirects revenues to the poor.

Reducing pollution that imposes high costs on cities

The effects of pollution on human health and productivity can be extremely high, especially for airborne particulate, lead (from all sources), and pathogens spread by water and food. These problems seem to be ubiquitous, at least in the larger cities. Priorities for action have to be related not only to the impacts of environmental problems, but also to the

84

costs and effectiveness of possible interventions: low-cost solutions are must needed in the developing countries. Bringing basic environmental services, top among them clean water and basic sanitation, to all city dwellers must be done as advocated in the previous paragraph. In addition, city-wide solutions such as trunk sewers, treatment plants and landfills will eventually be needed - if the right design choices are made and phased in over 15-20 years, this measure is affordable. The overwhelming issue is how to improve environmental management, both within sectors and across sectors and jurisdictions with urban areas. Local communities and polluters must be involved in developing solutions so as to build social and institutional sustainability. And the right incentives need to be put in place, and distorting subsidies removed, to achieve economic and financial sustainability.

Building sustainable institutions for managing the urban environment

The developing countries need to build up or strengthen their institutions. As the complexity of environmental problems grows with city size and economic development, so too should the capacity to respond grow. The focus initially should be on the simple, immediate priority interventions that can succeed and will lay the groundwork for solving future environmental problems. Local authorities need to find allies to build capacity for better environmental governance; deal with environmental spillovers through cooperation with neighboring municipalities and metropolitan, regional and national authorities; and build coalitions with communities, the private sector and NGOs for effective participation in neighborhood improvements. Finally, institutions should avoid excessive reliance on regulations and look for instruments that will change behavior, relieve conflicts and encourage cooperative arrangements.

Towards urban environmental management strategies

The World Bank's urban projects reach thousands of cities in the developing world. Since the 1992 "Earth Summit" in Rio, the Bank's environmental lending for "brown" issues such as urban water and sanitation, air and water pollution, solid waste disposal and energy has totaled 55 projects in 30 countries, involving Bank commitments of nearly US$ 6 billion. As part of this portfolio, support is being provided to a growing number of cities for formulating environmental management strategies and action plans, implementing investment programs, and carrying out critical policy and institutional reforms. In this work, emphasis is being put on a new policy framework for urban environmental management that is more strategic in nature; more oriented towards the wants of citizens and involving them in solutions; more focused on market incentives than regulation; and more visionary regarding the role of the private sector, both formal and informal.

Building on the experience acquired through some 25 years of urban lending totaling over US$ 25 billion - as well as benefiting from partnerships with others such as the Urban Management Program, the Metropolitan Environmental Improvement Program, the Sustainable Cities Program, and the Water and Sanitation Program - the Bank has sought to enunciate the new framework. in a series of policy papers. The first was *Urban Policy and Economic Development: An Agenda for the 1990s* (Cohen, 1991), and later papers included *Housing: Enabling Markets to Work* (Mayo and Angel, 1993) and *Better Urban Services: Finding the Right Incentives* (Dillinger, 1995). In the paper *Toward Environmental Strategies for Cities: Policy Considerations for Urban Environmental Management in Developing Countries* (Bartone et al, 1994), a strategic planning approach that is increasingly being adopted in the Bank and other donor-supported projects is summarized.

The strategic approach to urban environmental planning and management is based on enabling participation and building commitment. It has been tested in cities in industrialized and developing countries and is a viable approach for cities working towards setting up local versions of Agenda 21. The approach can involve several activities, each of which should emphasize strengthening local capacity, such as:

1   Informed consultation in which rapid assessments are conducted, environmental issues are clarified, key stakeholders are drawn in, political commitment is achieved, and priorities are set through an informed consultation process.

2   The formulation of an integrated urban environmental management strategy that embodies long-term goals and phased targets for meeting the goals; and agreement on issues-oriented strategies (that cut across the concerns of various stakeholders) and actor-specific action plans (that cut across various issues for achieving the targets, including the identification of least-cost project options, policy reforms, and institutional actions.

3   Follow up and consolidation in which agreed programs and projects are initiated, policy reforms and institutional arrangements are solidified, the overall process is made routine, and monitoring and evaluation procedures are put in place.

Latin America provides several examples of urban environmental management strategies applied in several different contexts. At the national level, the Colombia Urban Environmental Management Project is a Bank-supported initiative of the Ministry of Environment, approved in 1995, with the objective of strengthening environmental management institutions in the urban centers of Barranquilla, Bogota, Cali and Medellin, as well as by promoting environmental planning in selected medium size cities. The project will assist with the preparation of environmental management strategies and action plans in each of the cities. It is expected that it will be followed by investment projects aimed at implementing the action plans of the individual cities.

At the metropolitan area level, issue-specific environmental management strategies have been developed in a variety of complex situations. For air pollution control, a good example is the World Bank-supported Transport Air Quality Management Project for the Mexico City Metropolitan Area which was approved in 1994. As described in Bartone et al (1994), the project includes actions in the areas of vehicle emission control technology, alternative fuels and vapor control, and a transport policy and management component that focuses on travel demand management, urban freight management, public transport, and transport investment planning. It also aims at strengthening institutional capacity and scientific capacity for air pollution control

In Brazil, three major water pollution projects have been approved in recent years for the metropolitan regions of São Paulo, Curitiba and Belo Horizonte. These projects attack water pollution from both point and diffuse sources, the latter introducing complex land use issues. The São Paulo case is described in detail below.

Finally, in relation to basic services for the urban poor, a number of projects in Latin America, Africa and Asia involve the application of strategic sanitation planning. An example from Africa is provided below to illustrate how it is applied - the Kumasi Strategic Sanitation Plan.

Catchment protection in the São Paulo Metropolitan Region

The formulation of a watershed protection plan for the São Paulo Metropolitan Area (SPMA) is a good example of an issue-specific urban environmental management strategy. It started with a rapid assessment to gauge explicit needs and inventory problems. Consensus among the various stakeholders was achieved through consultations, culminating in an issue-specific strategy. The plan set the stage for coordinated short- and medium-term sectoral action plans and investments.

In the São Paulo approach, once priorities were set and consensus was achieved, a series of sectoral action plans and investments were designed. The lesson from São Paulo is that such a consensus-based strategy should comprise the following: the agreed long-term environmental goals for the urban region; a set of interim environmental goals and objectives to guide phased investments; the ranking of pollution control and other measures to improve environmental quality; the identification of priority sectors for channeling investments, including project profiles; and the recommended policy reforms, instruments, and institutional development needed to implement the environmental management strategy.

Background

The greater São Paulo Metropolitan Area (SPMA) is the most urbanized, industrialized and affluent region in Brazil. It consists of 38 cities, with São Paulo City (SPC) being the largest (current population is 11.4 million). The SPMA is already one of the largest and fastest-growing urban regions in the world, having a population approaching 20 million. From a growth rate averaging nearly five per cent annually from 1960 to 1980, it has slowed to 1.9 per cent annual growth in the last decades - 1.2 per cent in SPC and 3.2 per cent in the periphery. The SPMA covers 8,051 km^2 including a metropolitan core of 900 km^2. The region, with 12 per cent of Brazil's population, produces nearly a third of the country's industrial output and accounts for 18 per cent of gross domestic product (Leitmann, 1994).

Further expansion of the SPMA is bounded on the North and South by major water basins which are the sources for most of the urban water supply. Urban growth, however, has been in a Southerly direction resulting in water quality deterioration in the Guarapiranga Reservoir - a major source that serves 22 per cent of the urban population. Population in the Guarapiranga Basin is growing at seven per cent per year and has reached 580,000 inhabitants, of which two-thirds are located around the reservoir itself, and much of the remainder located in three adjacent municipalities in the upper catchment area. Some 24,500 families, however, live in *favelas* without urban services.

The wastewater from the adjacent communities are discharged directly into the reservoir along with their solid wastes. Upstream municipalities are sewered and have solid waste collection services, but are prohibited by federal and State law from discharging wastes within the catchment area. Compliance is impossible as neighboring municipalities outside of the catchment area have passed local bylaws prohibiting the importation of wastes from other jurisdictions. Federal and State land use restrictions in the catchment area are so severe that landowners do not develop their properties leaving them vulnerable to invasion by squatters to form new *favelas*. Agricultural and industrial activities in the catchment area are also problematic and better controls are needed.

Water quality of the Guarapiranga reservoir has been seriously compromised as a result of these problems. For example, over the past decade nutrient discharges (nitrogen and phosphorus) have increased resulting in the proliferation of algae in the reservoir and occasional fish kills. This in turn has led to a doubling of treatment costs, obstruction of treatment filters, and taste and odor problems. The water quality problems are well publicized and widely recognized by the urban population. In a city-wide environmental forum in 1991, the Guarapiranga Reservoir was unanimously cited as a priority issue requiring urgent action.

The São Paulo consultation

Broad recognition of the concern for water quality problems emerged from an initial consultation process sponsored by the Mayor of São Paulo and carried out in 1991 with Canadian support. To back up that concern with hard data, academic and NGO consultants were engaged for the preparation of a rapid environmental assessment and an environmental profile of São Paulo (Leitmann, 1994). The profile, which included interviews with city, community, and business leaders, was followed up by a town forum on the urban environment. Participants included over 120 city and State government officials, members of NGOs and community groups, academics and researchers, leaders of professional and business organizations, and members of the press. Among the priority issues flagged in relation to water quality in the Guarapiranga watershed were substandard housing, lack of environmental infrastructure and services for the poor, and settlement on risk-prone areas.

As an aside, it is important to note that the consultation process initiated in São Paulo has spread to many other cities. In 1994, a national consultation on urban environmental infrastructure and services included the preparation of environmental profiles and town forums in the nine metropolitan regions and five other important urban centers. The national consultation culminated in seminar in Brasília including the 14 participating cities, federal and State agencies, and professional associations and NGOs. The seminar resulted in major recommendations for the modernization of the environmental services sector in Brazil. More recently, eight cities have completed the preparation of Local Agenda 21's, following the recommendations of the 1992 Rio Summit - among them, São Paulo (PMSP, 1996).

As a result of the growing severity and awareness of the water quality problems in Guarapiranga, a working group was set up in early 1991 to come up with a solution. The group, led by the State Secretary of Water Resources, Sanitation and Works (SRHSO), included among others the State Water and Sanitation Company (SABESP), the State Secretary of the Environment (SMA), the Metropolitan Planning Authority (EMPLASA), the Housing and Urban Development Company (CDHU), the Electricity Company of São Paulo (ELECTROPAULO), and the four municipalities involved in the Guarapiranga Basin. An environmental strategy was formulated for the basin and an action plan prepared and presented to the World Bank for financing - which was approved in 1993. The Bank participated in and supported all phases of this process. The action plan is estimated to cost US$ 262 million, and includes the elements discussed below.

Immediate urban areas Factors to be considered include land use rationalization and control, including lot readjustment and urban rehabilitation; slum upgrading with some resettlement of low-income households located in risk-prone and geologically unstable areas; access roads and urban infrastructure; integrated provision of water supply, sewerage, drainage and solid waste collection in presently unserved areas; and elimination of direct waste discharges and runoff into the reservoir by interception and pumping to an adjacent basin (the Tiete River) where wastewater treatment will be provided.

Upper catchment area Factors to be considered include facilities for upstream wastewater treatment and solid waste disposal; land use management; development of recreational parks and ecological areas as buffer zones around the reservoir shoreline; control of erosion and nutrients from agricultural runoff; and restrictions on mining and industrial activities.

Responsibility for these actions is spread across several state and municipal authorities, as listed earlier, each of which has made a commitment to participate and finance their share of the agreed actions. Key instruments to be applied so as to meet the objectives of the project and ensure the desired benefits include: a surcharge for Guarapiranga water use by São Paulo consumers; adequate pricing policies for resources and services in the catchment area; increased property taxes and application of betterment levies to all local beneficiaries; development of recreational parks to protect critical and damaged areas and prevent future occupation by favelas, managed by private concessions; integrated pollution control actions across state agencies and municipalities; training programs for watershed management, including technical and financial support to NGOs and the four municipalities; and design and implementation of a basin authority.

Conclusion

One of the main medium-term institutional reforms will be the creation of a Guarapiranga water basin management agency. The action plan calls for studies and the organization of the public and governmental consultation processes needed to develop and establish the basin agency. An indirect benefit of the project will be to develop a modern scheme of river basin management, which, in the medium term, would be expanded on a larger scale to the other major basins in the region that serve the SPMA, such as the Piracicaba and Tiete Basins, as well as nationally.

The Strategic Sanitation Plan for Kumasi, Ghana

Kumasi has had three master plans in the last 40 years, but still has no comprehensive sewerage system and sanitary conditions continue to deteriorate as the population grows. The residents of Kumasi already pay about US$ 1 million per year to have only 10 per cent of their waste removed from their immediate environment. The current system of human waste management in Kumasi is inadequate; most of the waste removed from public and bucket latrines end up in nearby streams and in vacant lots within the city limits creating an environment prone to the spread of disease. With increasing rapid urbanization and competition for limited resources, there is the fear that the already poor sanitary conditions will worsen if no urgent and rational actions are taken.

In response to the inadequate sanitation conditions prevailing in the city, the Kumasi Metropolitan Area Waste Management Unit, with the assistance of the UNDP/World Bank Regional Water and Sanitation Group for West Africa, prepared a Strategic Sanitation Plan (SSP). The SSP reflects the willingness of the Kumasi Metropolitan Assembly (1993) to take the institutional and financial actions needed to ensure delivery of affordable sanitation service to all segments of the population by the year 2000. The plan differs from a traditional master plan in that it tailors recommended technical options to each type of housing in the city; considers user preferences and willingness-to-pay; uses a relatively short planning horizon (10-15 years), emphasizing actions that can be taken now; and breaks the overall plan into projects that can be implemented independently but which together provide full coverage.

The SSP moves away from reliance on conventional sewerage alone, and considers a range of proven technologies which address the needs of all segments of the urban population, recognizing resource constraints, and paying due attention to the willingness and capacity of users to pay for improved services (Whittington et al, 1992).

Background

The Kumasi metropolitan area covers some 150 km^2 and is made up of four districts. With a population of 575,000 in 1990, the metropolitan area is expected to grow to over one million population by 2010. For the purposes of sanitation planning in Kumasi, four major types of housing areas were identified:

1 In the areas of tenement housing (high density, low income), most people live in one or two room apartments in two- to three-story buildings shared by 10 to 20 families (40-100 persons per building). The buildings tend to be built around the edges of properties and have a central paved patio. Access is by the front street, with a passageway out the back. Population density varies from 300 to 600 persons per ha.
2 In the indigenous housing areas (low density, middle income), residences are generally one-story buildings with five to 10 rooms shared by four to 10 families (20-50 persons). The floor plans of these buildings is similar to the tenements. Population density varies from 80 to 250 persons per ha.
3 In the new estate housing areas (medium density, middle income), homes are generally one-story bungalows built in rows. Each home is occupied by one or two families, with a population density of about 50 persons per ha.
4 In the areas of high cost housing (low density, upper income), homes are generally separate one-family structures on large lots. Population density is about 10 to 15 persons per ha.

The Kumasi Metropolitan Assembly (KMA) is the municipal authority responsible for general city management, comprising four submetropolitan councils and 60 electoral areas. The administration of sanitation services in the KMA area was fragmented across three agencies - the Medical Department, the Mechanical Engineering Department, and the Metropolitan Engineering Department - with overlapping responsibilities for planning, development, and operation and maintenance of sanitation and solid waste services. Close to 50 per cent of municipal expenditures were spent on these activities. Typical sanitation services included:

1 Bucket latrines: used by 150,000 persons, and emptied three-five times per week by some 500 workers with little supervision.
2 Septic tanks with water closets: used by about 150,000 persons, including all high cost housing, 65 per cent of new estate housing, and 15 per cent of the tenements. Many septic tanks are in areas with insufficient space for infiltration fields, and often overflow into surface drainage.
3 Sewers: two small scale sewer systems service some 6,000 persons in government and commercial centers. Both discharge without treatment.
4 Public latrines: almost 260,000 are served by neighborhood public latrines, including water closets (60 per cent), bucket latrines (25 per cent) and ventilated improved pit (VIP) latrines (15 per cent). Public latrines charge about five cents a visit, but these funds have not always been used to maintain the latrines.

The most urgent problem is the removal of human wastes from the public latrines, bucket latrines, and septic tanks. About 70 per cent of the excrement is not removed from the urban environment. Only three vacuum tankers were available to service the city, but a recent ODA donation of five more vacuum trucks has improved the situation. If operated efficiently, these trucks would have sufficient capacity to collect and transport human wastes from the city.

Assessing the residential sanitation options

Sanitation planning was based on a combination of technologies and service levels. On-site options included VIP latrines, pour-flush toilets and septic tanks. Sewer options included conventional sewers, simplified sewers and small bore sewers. Each option was subject to technical, financial and social evaluation.

Technical considerations included the type of housing, water supply system, geologic conditions and operation and maintenance requirements. In tenement areas, only sewers were viable given the lack of space for septic tank drainfields (typically about 50 m^2), and lack of ventilation for VIP latrines. In the indigenous areas, VIP latrines, pour flush latrines, and septic tanks with water closets were all viable but sewers were not feasible given that only 25 per cent of homes in these areas have piped water. In the estate housing and high cost areas, all options were technically viable.

Main financial considerations included construction costs, operation and maintenance costs, and willingness to pay. For estimating the latter, contingent valuation surveys were conducted. The study indicated that people were willing to pay about as much for a KVIP latrine as for a water closet connected to a sewer. Conventional sewerage would not be possible for most households without massive government subsidies, whereas only modest ones were required for KVIPs. The demand-driven process thus helped determine which technology people wanted and which they and the government could pay for.

Social preferences were also determined by surveys. The surveys found that different groups have marked preferences with good reasons. Among the factors determining preferences were: cleanliness; privacy and convenience of the option; whether or not piped water was available; cost of water; and whether or not investments had already been made.

The Strategic Sanitation Plan

The SSP proposes that simplified sewers would be used in tenement areas, latrines or septic tanks in the indigenous housing areas, and septic tanks in estate and high cost housing areas. The city would subsidize sewerage in the tenement areas and latrines in the indigenous housing areas. To supplement the residential installations, the city would finance the construction and rehabilitation of public latrines in select areas and commercial centers.

Sewers Simplified sewers would be built for about 25 per cent of the population. This would include 150 km of sewers, 4,000 connections, and 16 ha of waste stabilization ponds, for a cost of US$ 10 million. Cost estimates are realistic, based on a pilot project financed by the UNDP and the KMA. The works are under construction and about half the target population was connected by the end of 1995.

It is proposed that a private contractor will operate and maintain the sewer systems and treatment ponds as well as collect user charges for the service. The contractor would pay a franchise fee to KMA sufficient to cover the debt service and finance the costs of the new Waste Management Unit. The sewerage tariff, though collected independently, would be based on monthly water bills and is expected to generate about US$ 1 million per year, a third of which would be paid to the city as the franchise fee. The contractor will be selected through competitive bidding - the sole criteria being the amount of the franchise fee to be paid to the city.

94

On-site systems Close to 350,000 persons living in indigenous housing areas will be served primarily by VIP latrines at an average cost of US$ 20 per person. Some 13,000 units will be needed each serving 25-30 persons. The cost of latrines will be shared 50-50 by the city and the beneficiaries. It is hoped that half of the homes eligible for a subsidy will construct a latrine within five years and that the city and beneficiaries will each pay about US$ 1-1.5 million. Local artisans and small entrepreneurs will be selected through a prequalification process and trained to market, design and build VIP latrines. The new sanitation department would approve designs and inspect construction on site, and would pay 25 per cent on approval and 25 per cent upon acceptance of the completed latrine. The city would also help develop the market through newspaper, radio and television advertising.

Public latrines would still be need to provide for about 100,000 persons. Construction, rehabilitation and operation and maintenance will be contracted out. Public latrine operators will pay a franchise fee to the city and are expected to have sufficient income to recover investment and operating costs. School latrine construction or rehabilitation is needed for 300 to 600 existing schools. The Waste Management Unit will work with parent teacher associations to develop plans for administering the latrines and educating users. Parent teacher associations will pay 10 per cent and 90 per cent will be subsidized by a Sanitation Fund (see below). The city will rent or sell its vacuum trucks to private operators who in turn will provide commercial cleansing services for septic tanks, VIP latrines and pubic latrines. Septage will be transported to the city landfill site where it will be co-composted with garbage to produce fertilizer.

Institutional arrangements In accordance with the SSP, the KMA will cease providing sanitation services directly and will instead enable the participation of the community and the private sector for providing these services. To facilitate this transformation in a transparent and responsible manner, the KMA established a Waste Management Unit (WMU) which in the final phase will be financially independent of the Treasury Department. The WMU will be responsible for the collection, treatment and disposal of human, industrial and solid waste. As part of its functions, the WMU will periodically update the SSP, mobilize resources for implementing it, supervise the planning and construction of facilities, supervise contract services, establish waste discharge regulations, and ensure compliance with the regulations. The WMU will be organized primarily to administer contracts, and will publish annual financial reports in accordance with commercial principles. It will have a board of directors made up of representatives of the KMA and the users.

As part of the transfer of services presently provided by the KMA to the private sector, its personnel was reduced from approximately 900 staff to about 100 by the end of 1994, and will be eventually reduced to 35.

The four submetropolitan districts of the KMA will continue to be responsible for administering public latrines in their respective areas. Each district will subcontract with the private sector to operate the private latrines and charge user fees.

The Sanitation Fund The financing of the SSP will depend on the Government of Ghana obtaining grants and credits which will be transferred as a subsidy to the KMA for the establishment of a Sanitation Fund. The Fund will be used for the construction of residential, school and public latrines. The franchise fees of private companies that administer the public latrines and vacuum tanker trucks will be deposited in separate accounts, and the revenues used to finance the WMU and construct additional installations.

Conclusion

The strategic planning process used in Kumasi is dynamic and the SSP itself will evolve as experience is gained. This interactive process began with a pilot project funded by UNDP in which the various technical, institutional, and financial issues that were proposed in the SSP were evaluated and refined. The pilot project is in fact the first phase of city-wide implementation supported by a World Bank-financed project. The Ghana Urban Sanitation Project, approved in 1995, provides a vehicle for extending the SSP process to Accra and several other Ghanaian cities. The SSP process is now spreading to other African countries and has already led to a similar project in Ougadougou, Burkina Faso.

References

Bartone, C. et al (1994) *Toward Environmental Strategies for Cities: Policy Considerations for Urban Environmental Management in Developing Countries*, World Bank Urban Management Program Policy Paper No. 18: Washington, DC.

Bartone, C. and Rodriguez, E. (1993) "Watershed Protection in the Sao Paulo Metropolitan Region: A Case Study of an Issue-Specific Urban Environmental Management Strategy", *Infrastructure Notes*, Urban Note No. UE-9, World Bank Transportation, Water and Urban Development Department.

Cohen, M. (1991) *Urban Policy and Economic Development: An Agenda for the 1990s*, World Bank Policy Paper: Washington, DC.

Dillinger, W. (1995) *Better Urban Services: Finding the Right Incentives*, World Bank Policy Paper: Washington, DC.

Kumasi Metropolitan Assembly (1993) *Strategic Sanitation Plan for Kumasi*. Draft report, January.

Leitmann, J. (1994) *Rapid Urban Environmental Assessment: Lesson from Cities of the Developing World*, World Bank Urban Management Program Discussion Papers Nos. 14-15: Washington, DC.

Mayo, S. and Angel, S. (1993) *Housing: Enabling Markets to Work*, World Bank Policy Paper: Washington, DC.

Prefeitura Municipal de São Paulo (1996) *Agenda 21 Local - Compromisso do Municipio de São Paulo*, Secretaria Municipal do Verde e do Meio Ambiente: São Paulo.

Roche, R. (1996) *Sanitation Planning in Kumasi*, World Bank Transportation, Water and Urban Development Department Course Notes: Washington, DC.

Whittington, D. et al (1992) *Household Demand for Improved Sanitation Services: A Case Study of Kumasi, Ghana*, UNDP/World Bank Water and Sanitation Program Report No. 3: Washington.

World Bank (1996) *Livable Cities for the 21st Century*, World Bank Environmentally Sustainable Development Vice Presidency: Washington, DC.

97

6 A new air pollution program for Mexico City

Leonardo Martínez-Flores

Introduction

This paper aims to present a summary of the principal components of the Air Pollution Program for the Metropolitan Area of the Valley of Mexico *(Programa para Mejorar la Calidad del Aire en el Valle de México 1995-2000)*, as well as to provide a clear presentation of the Program's philosophy.

In March 1996, the Mexican environmental authorities presented a new air pollution program for the Metropolitan Area of the Valley of Mexico (MAVM). The challenges of the program were enormous, not only because of the complexity of the problem itself, but also because of the social and political pressure from the metropolitan population, demanding lower levels of air pollution in the Valley of Mexico. Demands that are, undoubtedly, constantly reinforced by the publication of new information regarding the serious effects caused by pollutants on the population's health.

That is why the Program has the general purpose of protecting the population's health in the MAVM, by reducing air pollution levels on a gradual and permanent basis. It is important to note that the Program has its roots in a new conceptual framework approaching the air pollution problem from a systemic and integrated perspective, making use of the actual knowledge that we have today on environmental problems, relevant technologies and other national and international experiences. This new concept, it should be stressed, considers the urban phenomenon as a dynamic and open system, relating environmental quality to a number of vital urban processes such as: the spatial structure of the city (its size, population densities and land use diversification); the public and private systems of transportation; the road structure; the coexistence of multiple markets; the state-of-the-art of many different technologies; information systems; and the customs and traditions of the Mexican urban culture.

We were aware from the inception of the Program that the achievement of fundamental solutions depends on the induction of a profound and durable cultural change, capable of modifying the type of relation that we currently maintain with the city and the environment. In this context, it is important to favor a gradual and progressive change in the scheme of values and in people's priorities in order to match them to a realistic project of an urban sustainable development.

It is widely known that the air quality in the MAVM is very poor. However, the measures that, over several years, have been adopted in order to stop the degradation of air quality have already generated some important results: the previous rising levels of some air pollutants have been successfully controlled (such as lead, sulfur dioxide and carbon monoxide); Mexican gasolines meet international standards, with unleaded gasoline being superior to the average equivalent from the USA and most Asian and European countries; as to leaded gasoline, the lead component was reduced by 92 per cent in recent years. Diesel has also been improved by reducing sulfur by 95 per cent, and other widely used and highly pollutant industrial fuels have been gradually replaced by cleaner fuels as natural gas.

Unfortunately, we have also to acknowledge that some other air pollutants have reached unacceptable levels. This is the case of ozone and particle matters of less than 10 micrometers of aerodynamic diameter (PM10). In the particular case of ozone, in the last years it has reached levels that surpass the air quality standards in about 90 per cent of the time.

The structure of the Program

The Program has four general aims, namely: clean industry, through emission reduction per aggregated value in industry and services; clean vehicles, by means of lowering emissions per kilometer; efficient transportation and a new urban order, through the control of total kilometers traveled by all vehicles; and ecological recovery, including the combat of erosion.

These aims comprise several strategies, each one of them containing in themselves a certain number of policy instruments, amounting to a total of nine strategies and 94 instruments. The strategies are as follows: to include and permanently improve new industry and services technologies; include and permanently improve new vehicle technologies; improvement and substitution of energy sources in industry and services; improvement and substitution of energy sources for vehicles; wide provision of efficient public transportation; metropolitan policy integration (urban development, transport and the environment); economic incentives; inspection and enforcement of regulations in industry and transport sectors; and environmental education and information as well as citizens' participation in decision-making. The principal instruments of these strategies will be discussed later.

The long-term problem is how to choose a cost-efficient combination of strategies and instruments that favors the reduction of pollution levels per day, and the number of environmental emergencies per year. An environmental emergency, in this context, generally means a situation in which, on a particular day, the ozone level reaches 250 IMECA points, IMECA being the index used to measure the levels of pollutants. This means that the average probability distribution of the air quality index has to be pushed to the left, as shown in Figure 1.

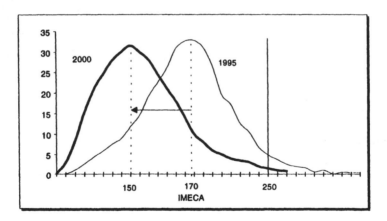

Figure 1   Frequency distribution of IMECA for 1995 and 2000
Source: Instituto Nacional de Ecología, Mexico City, 1995

Criteria used to define strategies and policy instruments

The whole pollution process begins with various users of the air basin of the Valley of Mexico, as they are the ones who most pollute the atmosphere. Basically, they are the industrial, commercial and services establishments, as well as private and public motorists. Until very recently, such agents made nearly unrestricted use of the natural environment of the air basin, with no other limitations than those of their own choice and possibilities. This situation can be easily compared to the so-called "tragedy of the commons", where each agent pursues their own interests and maximizes their personal benefits or profits, using without restrictions the air basin virtually to the point where their original benefits disappear.

However, the inherent costs in terms of degradation of the air quality are absorbed by the whole society, meaning that there is an uneven relation between the benefits - which are private - and the costs - which are public. In this typical situation, nobody adjusts unilaterally their actions for the sake of collective interests, because of two principal reasons, namely: the high individual costs implied, and the expectation that all others will not change their conduct. We have to overcome this typical problem of collective action: many agents will adopt the behavior of "free riders", doing nothing to pollute less, while they expect to enjoy, freely and without much effort, the benefits generated by the majority in terms of reduced pollution.

This is the reason why we think that a cultural change should be seen as a necessary condition to many of the proposed measures of the Program, because solutions are not without cost, and the determination to move forward necessarily implies that the inherent costs of their implementation must be in some way distributed among the real contributors to the problem. In other words, the willingness to pay for the proposed measures depends on several factors, one of them being of a cultural nature. It is needless to say that the distribution of costs must be made in a just and equitable manner. In this sense, it was evident to us that the scope of many of the proposals depends critically on the sense of co-responsibility and on the willingness to change the *status quo* of the metropolitan population.

However, there are two other crucial questions to be answered, namely: which activities should the strategies address and how the efforts and associated costs to reduce air pollution in the MAVM should be distributed. I will succinctly describe the principal aspects considered in the discussion of this topic. The first one has to do with the conceptual approach to the problem of air pollution in Mexico City. The second one is related to the relative contribution of pollutant emissions among the different sectors of urban activity.

The answer to the first question is schematized in Figure 2.

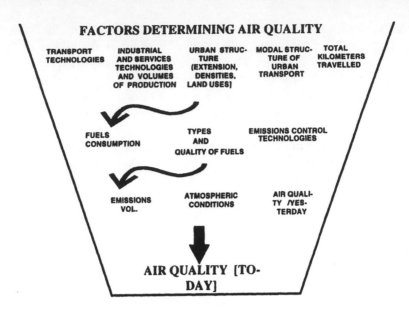

Figure 2 Factors determining air quality
Source: Instituto Nacional de Ecología, Mexico City, 1996

From this diagram, those variables that may and should be tackled by the strategies of the Program can be easily identified, as well as those variables which we have no possibilities of changing (for example, the atmospheric conditions and air quality of the day before). It should be stressed that, for the first time in Mexico, an air pollution program delves deep to identify and tackle the structural causes of pollution, abandoning the habit of designing end-of-pipe measures.

The second crucial aspect mentioned above is related to the relative contribution of pollutant emissions among the different sectors of the urban activity, which is closely related to the energy balance of all activities. The consumption of gasoline and diesel in the transport sector represents, at once, the largest relative energy expenditure and the largest percentage of pollutants emitted out of the total volume (including emissions of CO, NOx and HC).

The elaboration of an air pollutant emissions inventory is a strategic instrument for the management of the air basin. It reflects the intensity of use of the air basin capacity by the different sectors. It gives us an idea of the environmental efficiency of different urban processes as well as the priorities to be considered in designing programs and strategic measures. The development of a detailed, precise, and updated emissions inventory is a complex task that demands a systematic integration of information from several sources, with the active participation of federal and local authorities. Some international experiences also suggest that years of studies and a considerable amount of resources are necessary. The main results are organized in the Tables and Figures presented at the end of this paper.

Policy instruments of the Program

This is the type of information that allowed us to specify the policy instruments of the Program, which are listed below in a schematic manner.

Clean industry through emission reduction per aggregated value in industry and services:

New industry and services technologies
1 New standards for NOx, SO$_2$ and particle matters in industry.
2 Regional system of maximum limits of NOx emissions in industry.
3 New standards for VOCs.
4 New standards for natural gas industrial installations.
5 New standards for industrial and commercial use of liquid petroleum gas.
6 Self-regulation industrial program for environmental emergencies.
7 Technological conversion in the Jorge Luque and Valley of Mexico thermoeletrics to diminish NOx emissions.

Improvement and substitution of energy sources in industry and services
1 Final implementation of the vapor recovery system in distribution gasoline terminals.
2 New standards for industrial fuels.

Economic Incentives
1 New relative-price structure for commercial and industrial fuels, favoring cleaner fuels.
2 Fiscal incentives for emission control equipment and technologies.
3 Import tax exemptions to selected control equipment.
4 Credit lines to ecological industrial projects.

Inspection
1 Intensification of the Industrial Inspection Program.
2 Enforcement of regulations of vapor emissions in gasoline stations.

Environmental education and information and citizens' participation.
1 Pollutant Release and Transfer Register (PRTR).
2 Creation of the Clean Production Studies Center.
3 Permanent system of public evaluation of the Program with possibilities of incorporating innovative individual proposals.

Clean Vehicles: lowering emissions per kilometer

New vehicle technologies
1 Modification of the *Hoy No Circula* and *Doble Hoy No Circula* Programs (one-day and several-days vehicle circulation restrictions).
2 New standards for vapor emissions in circulating vehicles.
3 New standards for CO, HC and NOx in gasoline automobiles.
4 New emission standards for diesel vehicles.
5 New emission standards for new automobiles.
6 Updating the metropolitan data on all existing vehicles in circulation.
7 Improvement of vehicular verification (to include NOx, catalytic converter diagnosis, noise and dynamometer testing in gr/km).
8 Audit and permanent inspection of verification centers.
9 Homologation of verification process in all metropolitan area.

Improvement and substitution of energy sources for vehicles
1 New standards for gasoline.
2 Immediate reduction of toxic components in gasoline: Aromatics from 30 per cent to 25 per cent vol., max.; Olefines from 15 per cent to 10 per cent vol., max.; Benzene from two per cent to one per cent vol., max.; Vapor pressure Reid from 8.5 to 7.8 psi.
3 A Gas-Vehicles Development and Conversion Program.
4 An Electric Vehicles Industrialization Program.

Inspection
1 New street enforcement system for all types of vehicles.
2 Audit and inspection of vehicular verification centers.
3 Reinforcement of quality control system for gasoline.

Economic Incentives
1 Design of gasoline price policy internalizing social and environmental costs.
2 Annual vehicle tax modification internalizing environmental costs.
3 Change in relative prices of leaded an unleaded gasoline.
4 Tax incentives for new cars.
5 Reduction of import taxes to some gas engines and equipment.

Environmental education and information and citizen's participation
1 Identification of pollution levels on windscreens of cars
2 Permanent system of public evaluation of the program with innovative individual proposals.

Efficient transport and the new urban order through the control of total kilometers traveled by all vehicles

Wide provision of efficient public transport
1 New structure for surface public transport.
2 New regulations for public transport.
3 New regulations for freight transport.
4 System of executive buses in some avenues.
5 Construction of more kilometers of the Metropolitan Subway.
6 New lines for light electric trains.
7 Electric trains.
8 Extension of the trolley program.
9 Public transport only lanes in selected streets and avenues.
10 Better traffic management systems.
11 Organization and coordination mechanisms for taxis and minibuses.
12 Promotion of bicycle commuting.

Metropolitan-policy integration (urban development, transport and the environment)
1 Ecological land use zoning in metropolitan area.
2 Ecological areas protection program.
3 Diversification of land uses.
4 Recovery Program for the Historic Center of the City.
5 Restoration of urban zones.
6 Rationalization of new highways.
7 Integration of transport and urban development plans.

Economic Incentives
1 Design of urban economic instruments.

Inspection
1 Inspection system over the ecological areas.

Environmental education and information and citizens' participation
1 Bring up to date the simulation photochemical model IMP/Los Alamos.
2 New epidemiological studies.

3 Permanent system of public evaluation of the Program, with possibilities for making innovative individual proposals.

Table 1
Emissions distribution 1994 (tons/year)

Sector	Particles	SO$_2$	CO	NO$_2$	HC	Total	%
			Tons/year				
Industry (1)	6,358	26,051	8,696	31,520	33,099	105,724	3
Services (2)	1,077	7,217	948	5,339	398,433	413,014	10
Transport (3)	18,842	12,200	2,348.497	91,787	555,319	3,026.645	75
Vegetation and soils (4)	425,337	0	0	0	38,909	464,246	12
Total	451,614	45,468	2,358.141	128,646	1,025.760	4,009.629	100.0

Sources: (1) Instituto Nacional de Ecología, Sistema Nacional de Información de Fuentes Fijas, 1994; (2) Departamento del Distrito Federal, Dirección General de Ecología, Subdirección de Inventario de Emisiones y Atención a Contingencias, 1994; (3) Departamento del Distrito Federal, Subdirección General de Ecología, Dirección de Estudios y Proyectos Ambientales, 1994; (4) UNAM, Centro de Ciencias de la Atmósfera, *Reporte final de cálculos y mediciones de hidrocarburos naturales en el Valle de México*, 1994 and *Estudio de Emisión de Partículas Generadas por Fuentes Naturales*, 1990.

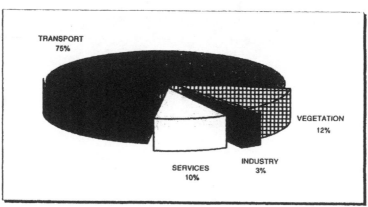

Figure 3 Emissions in the MAVM 1994 - Contribution to total emissions per sector of activity
Source: Instituto Nacional de Ecología, Mexico City, 1995

Table 2
Emissions distribution 1994 - pollutant percentage per sector

Sector	%				
	Particles	SO$_2$	CO	NO$_2$	HC
Industry (1)	1.4	57.3	0.4	24.5	3.2
Services (2)	0.2	15.9	0.1	4.2	38.9
Transport (3)	4.2	26.8	99.5	71.3	54.1
Vegetation and soils (4)	94.2	0.0	0.0	0.0	3.8
Total	100.00	100.00	100.00	100.00	100.00

Sources: (1) Instituto Nacional de Ecología, Sistema Nacional de Información de Fuentes Fijas, 1994; (2) Departamento del Distrito Federal, Dirección General de Ecología, Subdirección de Inventario de Emisiones y Atención a Contingencias, 1994; (3) Departamento del Distrito Federal, Subdirección General de Ecología, Dirección de Estudios y Proyectos Ambientales, 1994; (4) UNAM, Centro de Ciencias de la Atmósfera, *Reporte final de cálculos y mediciones de hidrocarburos naturales en el Valle de México*, 1994 and *Estudio de Emisión de Partículas Generadas por Fuentes Naturales*, 1990.

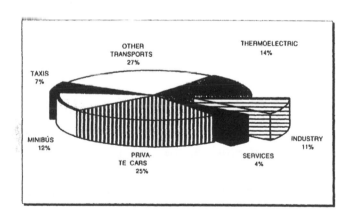

Figure 4 Emissions in the MAVM 1994 - Annual contribution of nitrogen oxides per type of source
Source: Instituto Nacional de Ecología, Mexico City, 1995

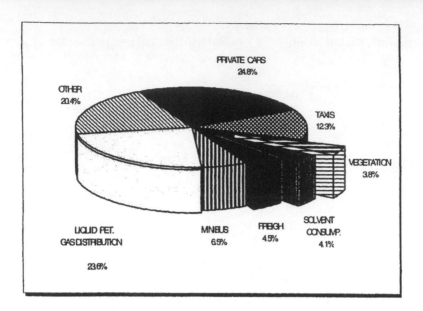

**Figure 5   Emissions distribution in MAVM 1994 - annual
            contribution of hydrocarbons per type of source**
Source: Instituto Nacional de Ecología, Mexico City, 1995

7 Towards sustainable mega-cities in Latin America and Africa

Janice Perlman

Introduction

By the year 2000, for the first time in history, more than half the world's population will live in cities. In twenty three of these cities, populations will exceed ten million. No precedent exists for feeding, sheltering, employing, or transporting so many people in so dense an area, under such severe financial and environmental constraints. Water has become undrinkable, air unbreathable, waste unmanageable, and open space unavailable. Cities are reaching the limits of their carrying capacity to support human life. Urban systems designed hundreds years ago for cities of a much smaller scale and seemingly inexhaustible natural resources are totally inadequate for the needs of the 21st Century.

Given the magnitude of urban environmental problems and the worldwide concentration of population and economic activities in cities, it is not surprising that many regard cities as the cause of global environmental deterioration. This misinformed perception, however, creates counter-productive policies which threaten to bring us to a point of no return. To reverse this downward cycle, we need a new paradigm based on the factual realities rather than an anti-urban bias.

Each day in our cities, we witness the traffic grow thicker, and the air dirtier. We observe the disappearance of parks and green spaces. And we stumble over the clutter of trash on our sidewalks. The media informs us that our water is undrinkable, that there are invisible toxins around us, and that disease rates are rising. The environmental problems of our cities have grown to such paralyzing proportions that there seems to be little we can do. In reality, there are many innovative local responses to these problems, but we do not hear much about them. Television reporters rarely cover them, newspaper journalists do not write about them very much, and the community organizations are too busy coping with survival and handling crises to undertake self-promotion. Consequently, many of these initiatives remain largely unknown, preventing others from replicating or adapting them.

The six points below may be controversial or appear counter-intuitive, but they are the inescapable basis for moving towards a sustainable and equitable society:

1 There can be no global ecological sustainability without urban ecological sustainability. Concentrating the human population in cities is an environmental necessity. Not only do the economies of scale create energy and resource efficiencies, but also, if the entire land mass of the planet were divided into individual household plots, there would be no space left for either agriculture or natural wilderness areas. Circular rather than linear systems are needed: as cities concentrate pollution and environmental degradation, transforming the urban metabolism through circular rather than linear systems is the key to reversing our global environment deterioration. We need to re-use our water and waste streams, and utilize what is now discarded as productive resource.

2 There can be no urban environment solution without alleviating poverty. The urban poor tend to occupy the most ecologically fragile areas of our cities, such as steep hillsides, low-lying swamplands, or adjacent to hazardous industries. In addition, their lack of resources often prohibits them from having adequate water, sewage, or solid waste management systems. Without alternative locations to settle and sufficient income to cook and keep warm, their survival will increasingly be pitted against environmental needs.

3 There can be no lasting solutions to poverty or environmental degradation without building on bottom-up, community-based innovations. Since creativity was not distributed along lines of race, class, or gender, experts and policy makers are not always the best source of system-transforming innovations. The most creative and resource-efficient solutions to urban problems tend to emerge at the grassroots level, closest to the problems being solved. And, without local participation in implementation, even the best ideas are doomed to failure.

4 There can be no impact of scale with sharing what works among leaders and scaling up into policy. While small may be beautiful, it is

still small. In order to have meaningful impact, micro-initiatives need to be replicated across neighborhoods and cities through peer-to-peer learning or incorporated into public policy frameworks.

5 There can be no urban transformation without changing the old incentive systems and "rules of the game". Since every sector of urban society holds a *de facto* veto on the others, local innovations can never achieve scale with cross-sectoral partnerships involving government, business, NGOs, academia, media, and grassroots groups. We need to create a climate conductive to experimentation, mutual learning, and collaboration.

6 There can be no sustainable city of the 21st Century without social justice and political participation, as well as economic vitality and ecological regeneration: innovations that promote ecological regeneration are just one piece of the puzzle. It is only by seeking out innovations in social equity, participatory democracy, and economic productivity - and ideally, innovations that possess all four attributes - that our cities can be truly be sustainable for the 21st Century and beyond.

This paper focuses on three case studies that highlight effective solutions to local environmental/poverty problems; they are from: Cairo, Egypt; Rio de Janeiro, Brazil; and, Accra, Ghana.

Methodological approach

Finding what works

These innovations are just a few from a vast array of successful innovations that have been identified by our Mega-Cities teams worldwide. Each of our Project Coordinators, based at an urban research center, university, or non-profit organization has undertaken a systematic search for local success stories. To structure their search, we use a five-pronged approach:

1 By policy area: we created task forces by policy area, such as economic development, housing, or environmental issues.
2 By geographic neighborhood: we surveyed planning and community development directors in each geographic neighborhood.
3 By sector: we networked through our multi-sectoral Steering Committees (including letters from government, business, NGOs, grassroots groups, academia, and the media).

111

4   Through literature reviews: we conducted in-depth literature reviews of books, journals, magazines, and newspapers.
5   Through a "call for innovations": we utilized the media to make a "call for innovations", asking readers, listeners, or viewers to respond with initiatives in their neighborhood that have made life more workable.

For the purpose of this project, each innovation identified through this five-prong process was then screened according to three sets of criteria, namely value, impact, and operational:

1   Value criteria: in addition to being environmentally sustainable, each of the innovations needed to demonstrate economic viability, political participation, social equity, and cultural adaptability.
2   Impact criteria: the innovation needed to demonstrate significance, quality, scope, proven merit, evidence of lasting change, and above all, novelty, or "innovativeness".
3   Operational criteria: under this category, the innovation needed to be cost-effective, practical, collaborative, replicable, and sustainable. If it was a good idea, but was simply too expensive or difficult to implement, then it was removed from the list.

Through this process, the Mega-Cities teams worldwide have identified hundreds of successful urban innovations across three primary areas: environmental regeneration, poverty alleviation, and democratic participation. The environmental cases featured in this paper are just an illustrative sampling.

An array of approaches

The innovations described demonstrate a range of approaches, including urban gardening, reforestation, recycling, job training, elementary education, etc. Despite the rich diversity of these case studies, some common themes run through them. Several of the innovations, for example, represent attempts to rethink the way we build the streets, drainage systems, buildings, and other physical infrastructure of our cities in a more environmentally friendly manner. In Rio de Janeiro, the Paid Self-Help Reforestation Project organizes the residents of hillside squatter settlements to separate sewage and drainage systems and plant fruit trees as a way of reducing landslides and increasing nutrition.

Other innovations seek to redefine economic development in environmentally sustainable terms. In Accra, Ghana, the Urban Market

Gardens Project has enabled low-income residents to create vegetable gardens on unused land, filter waste water for irrigation, and earn income by selling their produce in vibrant public markets. In Cairo, Egypt, the Zabbaleen Program has provided informal sector trash collectors with the equipment, technical assistance, and start-up funds necessary to create their own micro-enterprises converting recyclables into marketable products. This fundamental shift in power "changes the rules of the game", and, if implemented widely, could profoundly alter the degree to which low-income communities can control their environmental future.

A third theme of these innovations is a focus on altering human behavior through environmental education and training. The power of these innovations lies in their recognition that every urban citizen, no matter how young or old, must understand the environmental consequences of their actions, and be a part of the solution.

For comparability, each innovation has been documented in a common format: the background and context; project description; obstacles; impact; scaling up the impact.

Preliminary findings

Findings about the environment/poverty issue:
1 The Environmental Justice Movement (fighting the concentration of environmental hazards in low-income communities) applies equally in the North and the South.
2 Because health hazards permeate the rest of society, environmental issues can mobilize broader alliances than other social issues.

Findings about innovation origins:
1 In every case a "window of opportunity" sparked the innovation; either a discreet crisis or one of culminating deprivation.
2 Non-governmental organizations (NGOs), acting in the space between the market and the state, were an essential ingredient for success, whether as initiators, brokers, or implementors.

Findings about implementation:
1 Partnership is essential. There was no case of successful innovation without cross-sectoral partnership. The single case which failed lacked this partnership element, and could possibly have succeeded through more significant collaboration with the community.
2 Need product champion on both sides of the partnership.
3 Role of women is critical, showing up in case after case.

4 Conflict between need to show rapid results and drawn out consultative process.
5 Evolution from single issue to holistic approach.

Findings about obstacles:
1 Scarcity of resources: inadequate funds, land, technical assistance, information, etc.
2 Community problems: difficulty in achieving consensus, lack of cooperation among competing factions in the community, threat by gangs and drug-lords, burn-out, and inexperience with partnerships.
3 Government resistance: red tape, turf wars, fear of loss of control and power, fragmentation, legislation, and inexperience with partnerships.
4 Technological problems: rigid technology, costliness of testing new "appropriate technology".

Findings about impact:
1 On individuals: increased self confidence, environmental awareness and skills.
2 On stigmatized groups: new respect, new contributions.
3 On community organizations: capacity building.
4 On local government: changes in problem-solving approach and paradigm.
5 On political culture: building democracy, transparency, right-to-know, raised expectations, and inclusionary principles.
6 On poverty and environment: incremental steps can result in significant systemic change upon reaching a critical mass.

The Zabbaleen Environment and Development Program: garbage recycling and micro-enterprise development in Cairo, Egypt

Background and context

Cairo's population of fifteen million is growing at a rate of almost one million every eight months. The city's immense and explosive growth results in political, economic, social, and environmental imbalances that negatively affect the quality of life in the Greater Cairo urban region. The demand for services is great, often outpacing the ability of the government to provide them. One area of especially pressing concern is the collection and disposal of solid waste.

Greater Cairo generates around 6,000 tons of solid waste per day. Responsibility for the management of the solid waste system in Cairo is shared by the municipal sanitation force and by the traditional private sector waste collection system that has evolved within the last fifty years and which includes the Wahis and the Zabbaleen. The Wahis were the original garbage collectors, entering the trade after migrating to Cairo about 100 years ago. The Wahis later stopped collecting the garbage themselves, yet maintained their control of the garbage collection industry with the help of jobless and landless pig breeders who had moved to Cairo in search of work. To this day, the Wahis subcontract with the owners of buildings and serve as administrators of the system, while the migrants (now called "Zabbaleen", Arabic for garbage collector) collect the waste and transport it to their settlement, where it is sorted and recycled, or used for animal fodder.

In the 1970s, as the city continued to grow and the amount of daily garbage skyrocketed, Cairo's trash collection needs began to overtake the Zabbaleen's capacity to provide services. Furthermore, little of value was expected from the garbage produced by the majority of new immigrants who were poor, so there were no incentives to collect it. A marginalized community with little or no organization or power, the Zabbaleen lived in settlements with few basic services, and suffered from environmental devastation, little economic opportunity, lack of education, and a host of other problems endemic to urban slums. When the government threatened to look to more modern and efficient systems to dispose of the city's waste, the Zabbaleen faced the possible loss of their traditional livelihood.

Moqattam houses half of Cairo's garbage collectors and is the largest of the seven settlements of its kind. Prior to 1981, there was no water supply or sewage system, and very few of the homes had electricity. The settlement had no government school, health clinic, pharmacy, telephones, or means of transporting emergency patients to hospitals. The sorting of the refuse inside the houses left them cluttered and often filthy. The situation was mirrored by the condition of the roads, which were heaped high with waste paper, piles of animal manure mixed with organic residues, tin cans, and often animal carcasses. The waste had attracted countless flies, and the air was usually filled with the smoke of fires which were set deliberately to dispose of unwanted paper, or resulted from spontaneous combustion of organic residues. In the absence of a water supply on the site, fires destroyed large sections of the community on several occasions.

The primary objectives of the Zabbaleen Environment and Development Program were to improve the living conditions and build the capacity of the Zabbaleen in the Moqattam settlement, while creating a more efficient solid waste management system for Cairo. The program consisted of a number of projects initiated over a span of ten years, and was based on an exploratory, experimental approach. Activities were targeted at improving environmental and living conditions, promoting enterprise development, increasing the service capacity of the Zabbaleen, and instituting low cost technological innovations.

The program was first conceptualized by the Governor of Cairo and the World Bank in 1976. It was initiated and managed by Environmental Quality International (EQI), a Cairo-based consulting firm specializing in solid waste management and urban renewal, with the assistance of the Moqattam Garbage Collectors' Association, the Gameya. Numerous government and non-government organizations also participated in one or more aspects including the International Development Association of the World Bank, the Ford Foundation, Oxfam, the Soeur Emmanuelle Fund, and the European Community.

As an initial step, the Area Upgrading and Infrastructure Extension Project began in 1982, financed by the government of Egypt in cooperation with the International Development Association of the World Bank. EQI mapped out the settlement and, with the assistance of a consultant, the government constructed basic infrastructure and facilities, including piped water, electricity, sewerage networks, paved roads, a primary school and a health center. In an attempt to recover costs for the infrastructure and facilities construction and to provide residents with land security, community members were offered a 30-year-installment plan for the purchase of the land on which their families resided.

Work was then begun towards meeting the goal of an improved solid waste collection system for Cairo. The Route Extension Project was funded by Oxfam in 1982, to extend the Zabbaleen collection routes to 8,000 new homes in the neighboring low-income area. To create an incentive for collecting in areas which produce little value garbage, a fee-collection system was put in place.

In an effort to then diversify income generating possibilities, two projects were introduced. The Small Industries Project, funded by Oxfam, offered credit and technical expertise to small community-based recycling industries in order to maximize the resource value of waste. For example, the Small Industries Project enabled some Zabbaleen families to buy plastic granulating machines to recycle plastic and rag pulling machines to recycle rags. The Income Generating Project for Female-Headed Households, funded by the Ford Foundation, extended credit to widows, divorcees, and women with unemployed or disabled husbands. Beneficiaries decided what type of enterprise they would establish with their loan. Loan payment guarantees and support for borrowers was created through the formation of credit groups, following a tradition of collective saving groups.

In an effort to clean up the community while generating income, the Composting Plant was constructed in 1986, funded by the European Community, the Ford Foundation and the Soeur Emmanuelle Fund. Using simple rudimentary technology, organic waste in the settlement is removed, composted to make fertilizer, and sold to fund other development projects. The Association for the Protection of Environment (APE) was established to run the plant, and was eventually able to help initiate programs for women such as a school for rug weaving, literacy classes, a part recycling project and a health awareness program for pregnant women.

The Mechanization Project was initiated in 1987, when the Governorate of Cairo issued a decree banning the use of donkey-drawn carts for garbage collection. Funded by shareholders and supplemented from the Soeur Emmanuelle Fund, the Environmental Protection Company was established by EQI and the Gameya to provide vehicles for the settlement. Wahis and Zabbaleen organized themselves into more than 50 small private companies, mostly funded by the wealthier and more powerful Wahis. They operated in the neighborhoods they had both traditionally serviced. Working closely with their traditional clients enabled the Zabbaleen to introduce reasonable increases in prices that were sufficient to cover their traditional costs without loosing their client base.

Obstacles

Each of these and other projects experienced various difficulties and obstacles to success. Many of the projects and the program as a whole suffered from poor planning and a lack of coordination within the planning process. For example, the infrastructure system built in Moqattam brought basic services such as water almost solely to public areas. The majority of residents were not provided access to basic services, and, with the great increase in population that occurred, the services have failed to meet the needs of the community.

The pressure put on the system by high demand has resulted in the frequent breakdown of infrastructure components such as the sewer system. In addition, the overall *ad hoc* nature of the various programs resulted in a lack of coordination among the different organizations working in the community, preventing a cooperative, mutual learning approach. The large number of organizations and donors has also led to a habit of receiving hand-outs in the Zabbaleen community, in some instances creating dependency and limiting the chances for economic sustainability.

Financial and economic sustainability was another obstacle encountered in a number of projects, as experienced in the Income Generating Project for Female-Headed Households. Despite the establishment of a revolving loan fund, the project continued to require outside assistance because the fund did not factor in administrative expenses. Overall, however, the Zabbaleen Environmental and Development Program has in large part overcome financial constraints and proven sustainable. The continuation of extensive income generating activities only initially funded with assistance from the program demonstrates the sustainable improvements made.

Perhaps one of the most difficult obstacles to overcome has been the domination in participation and subsequent better access to program benefits by the richer, more powerful Zabbaleen. In the early years of the program, representation of community members in committees of the Gameya was in the range of 70-80 per cent, with outsiders constituting the remainder. As the years went by, however, representation from outside the settlement became a predominant feature while community representation became restricted to the more powerful families in the settlement. This led to self-interest driven decisions and the failure to reach the lowest income Zabbaleen. Coupled with the more willing nature of the rich to take a risk and try new ways, the project has increased the gap between rich and poor and exacerbated previous sources of conflict and tension. Nonetheless, some projects were able to address the issue by targeting benefits to the more vulnerable, as seen in the Income Generating Project for Female-Headed Households.

Impact

The Zabbaleen Environment and Development Program has been considered very successful in fulfilling its primary goals of improving the solid waste collection system in Cairo and improving the lives of the Zabbaleen community. Gauging the full impact of the Zabbaleen Environmental and Development Program is difficult; the overall indirect impact by far exceeds the direct impact of the program's individual components. The very fact of intervening in the community has activated a process of change that has its own dynamics. For example, a tile factory, several carpentry and blacksmith workshops, and the rug weaving and paper recycling industries have grown out of the program. Other development activities were also simultaneously inspired, such as the services provided by the local Coptic Church, the school established by the Order of the Daughters of St. Mary, and the polyclinic, literacy classes, sewing classes and nursery provided by the Integrated Care Society. These efforts have helped transform the settlement.

Direct results also brought obvious benefits. The most visible transformation in the settlement is its physical appearance. There are now approximately 1,500 houses in the settlement, many of which are multi-story structures, as compared to the existence of just over 700 one-story shacks in 1981. Most of the houses are built using concrete bricks with proper foundations. The number of inhabitants has almost tripled over 12 years, rising from 5,881 to 16,600 in 1993.

The cleanliness of the settlement has also improved due to the combined impact of the new infrastructure, internal clean up projects, and the composting plant. Health and hygiene have been aided through better health services and more sanitary means of collecting and sorting garbage, including the use of trucks instead of donkey carts. The health status of the community as a whole has improved, most notably in the decrease of infant and child mortality from 240 per thousand in 1979 to 117 per thousand in 1991.

Environmental benefits are numerous. The increased capacity of the Zabbaleen to collect garbage throughout Cairo has created a much cleaner city overall, in both higher- and lower-income neighborhoods. The recycling of organic and inorganic waste has significantly reduced the environmental burden of disposing it. The compost plant produces a product that is free of chemicals and harmful contaminants.

Economic benefits are also countless. The emphasis placed on institutionalizing the Zabbaleen trade has maximized the productivity and revenue generation potential of community residents. Household income has increased twenty times over the past ten years. Recycling activities and projects created a diversified urban economy and additional income. One of the main benefits of the shift away from garbage collection has been to women and children. It has relieved them of the long and arduous process of sorting, and has allowed them to engage in various other income generating, educational and recreational activities.

Finally, the program led to organizational capacity building and human development. It stimulated voluntary action through the Zabbaleen Gameya and encouraged the participation of community residents in implementation. Gameya staff were trained in program management, administration and extension work. Residents have increased their productive capacity through the use of the simple and appropriate technology promoted by the program. Educational opportunities have improved dramatically for the children of the Zabbaleen settlement. As a result of introducing motorized vehicles to replace donkey-drawn carts, child labor has been reduced, affording children the opportunity to pursue their education. School enrollment has greatly increased.

Scaling up the impact

The Zabbaleen Environment and Development Program can serve as a model for other cities seeking to solve problems of solid waste, without creating technologically complex and financially draining systems. By working with the Zabbaleen to enhance their traditional service capabilities and extend their routes, various agencies were able to meet a modern city's waste collection needs without installing a complex and expensive new system, and without displacing a group of traditional service providers. Through the UN Urban Management Program, aspects of this project are being transferred to other cities in the Middle East. Also, through the Mega-Cities Project components of the project are currently being transferred to Manila, Bombay, and Los Angeles.

As in these cities, several necessary conditions must be first satisfied before a transfer of the program can be attempted. In order to replicate the Zabbaleen Environmental and Development Program, there must already be a group that is providing waste disposal service, as the whole project is an attempt to legitimize and improve a traditional service sector industry and increase the capacity. This group must be willing to participate in the project. Also, the government must recognize the worth and potential of the traditional garbage collecting community and refrain from pursuing external, highly technological solutions to urban waste removal problems. Governments must further be willing to work with the local community and various NGOs and other agencies. This requires a considerable commitment of time, energy, political capital and financial resources.

As a project designed to increase the capacity of a traditional industry, a large element of community and human resource development must be included. Obviously, in order for a community like the Zabbaleen to enlarge their capacity, they must have access to basic amenities such as health-care, education, suitable infrastructure, water, sewage and others. Such extensive community development mandates the presence of organizations with experience in solving the technical and human resource problems of providing services to previously unserviced communities.

Finally and most importantly, any project based on the Zabbaleen Environmental and Development Program demands that the above actors, traditional waste collectors, and government and development agencies adopt a new attitude towards the informal trash collection industry and its potential in serving the needs of large, modern cities. This entails a basic and necessary commitment to several essential programs. Actors must be committed to the development of a decentralized, private recycling industry based on small-scale enterprises. In order to develop such enterprises, garbage collectors need assistance in developing entrepreneurial and business skills. Lastly, all actors must be willing to work as partners to develop the organization necessary to expand and upgrade traditional services and increase productivity.

The Paid Self-Help Reforestation Project in Rio de Janeiro, Brazil

Background and context

As home to 11 million people and the second largest metropolitan population in Brazil, Rio de Janeiro shares many of the problems of the world's largest cities. Lacking suitable planning structures, it has become the victim of uncontrolled development and overly rapid expansion, outgrowing its ability to provide basic services and putting unbearable strains on the urban environment. The poor inhabitants of Rio, many of whom live in informal squatters called *favelas*, bear the brunt of the problems.

The city of Rio has a long history of housing problems. The expulsion of slum residents at the turn of the 20th century and the absence of public policies to address the housing problem led to the first informal settlements on the hillsides of the city next to sources of work, in areas deforested by agriculture or logging. Unlike the more planned and regulated growth in the wealthier sections of the city, these favelas grew outside any formal planning structures and were deprived of municipal services. They became associated with poverty and stereotyped notions of marginality.

Over the following years, as the favelas spread their "incompatibility" with the suburban neighborhoods was widely recognized to the extent that they became viewed as intolerable. In the 1960s, the government attempted to forcibly evict favela residents from their homes, basing their policy on the premise that the favelas occupied land that was too good for them. Residents of the favelas, backed by public support, organized to oppose the evictions. The state finally recognized the favelas as part of urban reality and began to make efforts to improve the conditions. However, programs were generally large and government run, making few efforts to involve the community in improvement efforts.

Despite improvement efforts favelas still lack many basic services and opportunities, and residents suffer poor living conditions. In 1988, over 30 per cent of the poor in the metropolitan area were unconnected to the sewage system; over 35 per cent did not have piped water and over 50 per cent did not have their garbage collected. Unemployment is still high, and those that are employed generally have low-paying, unskilled jobs, often in the informal economy, lacking labor contracts and social benefits.

In addition, the favelas experience continued environmental degradation. Historically, Rio's many hillsides were the sides of coffee plantations and timber logging. The rapid growth of the favelas on top of the already exhausted soil only perpetuated the poor conditions: water saturated soil led to landslides and rock-falls; compacted soil caused excessive runoff and flooding; and the ever-present mud became the habitat of disease-carrying mosquitoes. Every year flooding kills hundreds of favela residents, and eroded soil destroys the city's drainage system.   24

Project description   24

In response to this problem, the Municipal Secretariat of Social Development (SMDS) created an integrated program to reverse urban deforestation and improve the social-economic conditions of the low income settlements. In collaboration with community groups and technical advisors, SMDS staff designed a scheme that addressed drainage, sewage, reforestation and environmental education in the favelas. Key to the project was the use of paid local labor, recruited and managed by community leaders, supported by a simple structure requiring few complex alliances and minimizing political maneuvering.

The Paid Self-Help Reforestation Program addressed several issues. While reducing the risks of flood and erosion from heavy rainfall and lowering the favelas' costly impact on city management, fruit from the planted trees improved residents' diets, and the local community members learned about the benefits of protecting the environment.

In 1979, the SMDS was set up to be responsible for transforming favelas through infrastructure and upgrading projects, coordinating efforts by government and community organizations. Out of this grew the

Self-Help Project, with reforestation undertaken in conjunction with activities such as road paving, drainage and sewage network construction and the building of health clinics and day care centers. The SMDS strategy for working with communities draws upon the traditional self-help networks that are an informal survival mechanism within the favelas.

The Paid Self-Help Reforestation Project in particular was created in 1986 following two requests. The municipal body in charge of slope stabilization and risk prevention asked that the municipal authorities involve the community in reforestation efforts to assist their work. At the same time, the Forest Engineers Union demanded that a greater role be given to their profession in local government.

The three-stage Self-Help Reforestation Project was implemented over three to four year periods, depending on the site. In the first stage, the SMDS would select a suitable site based on the following criteria: deforested hillsides with steep slope angles that presented a risk to the residents of low-income settlements; areas that included water sources and springs supplying rivers and drainage canals; and the presence of a well organized residents' association. Using hydrological characteristic maps, the SMDS engineers determined the appropriate growing conditions that would guide technical and financial plans.

The final part of the first stage was arguably one of the most important. The projects and its benefits would be presented at assemblies of residents' associations and discussed by the communities. A project manager would be elected to be responsible for the recruitment of labor, control of materials and participation of the local population.

As part of the second stage, forestry engineering skills were provided by the Center for the Production of the Forest (*Centro de Produção de Essências Florestais*) to assist in the selection of the appropriate tree species. Seedlings were cultivated by the center's mentally and physically disabled or homeless and jobless. Meanwhile, a year-long process of site preparation would occur. Access roads and ditches or canals of five meters width were built as needed. The sites were than cleared, plowed and demarcated into lots according to land characteristics. Terraces for tree growth and fertilization from a municipal waste recycling plant were also provided by the Center to prepare the site.

Planting began in Stage Three according to a process of ecological succession. Primary and secondary species were selected to prepare the soil for more fragile ones, corresponding to four varieties: strong, element-resistant plants; indigenous varieties; adaptable exotic species from the nurseries; and fructiferous species. After the initial planting, two to three years of critical maintenance were carried out by members of the community being paid by SMDS. Duties included cleaning drainage canals and ditches, replanting, weeding the control of pests and diseases. Over time the recreated forest will become self-sustaining, and will require little or no maintenance.

Obstacles

Though an innovative attempt at reforestation, the Self-Help Reforestation Program was met with several obstacles. Despite community participation and education regarding the need for reforestation, the local community often required wood for uses such as cooking. This resulted in the cutting down of fairly newly planted trees. Also, arable land is of high value in the already overcrowded favelas, which is another incentive to keep cutting down trees. Local herders who want access to the forest sites are against reforestation efforts and the threat to these areas from herds of grazing sheep, goats and cows is considerable. Compounding these was the weakness of SMDS's information campaign which was to promote the validity of natural resource management within the community. Without a strong educational component, the trees stand little chance of remaining untouched.

An institutional and legal obstacle which had to be overcome, and remains an obstacle to the transfer of the innovation to other areas of Brazil, is the legislation for the contracting and payment of labor. Under Brazilian legislation, the state can contract private labor for certain periods of time and for certain activities. To employ individuals, a public competition must be held that requires applicants to have reached a certain education level and that subsequently guarantees the employee a job for life. Intense political negotiation was required to make possible the contracting of local labor so that neither of these labor categories apply to requirements of the reforestation project. An alternative approach could be the use of NGOs to act as intermediaries for the handling of the finances and contracts.

A lack of political will of civic leaders toward financing such ventures also occurred, and project supporters had to continually justify the need for funding. A lack of intergovernmental coordination and NGO or community support, at one point or another, has also threatened the program's strength. To shield itself from such problems, the SMDS continues to maintain political isolation, but changes in municipal cooperation and political climate could weaken the institutional strength. In some cases where the participating community organizations did not accurately reflect the needs of the local community, project benefits were not appropriately administered. At times the SMDS technical staff also failed to respect local expertise, leading to friction.

Impact

As of 1994, the Self-Help Reforestation Program had directly affected 25 favelas, with 90,000 favela occupants and an impact on residents at the bottoms of the hills as well. Additionally, 300,000 seedlings have been planted in a total area of 241.8 hectares. In terms of the benefits from reforestation, erosion was reduced, thus lowering amounts of sediment flow into communities at the base of the hills. Previously overtaxed drainage systems, which had been blocked by silt, are now capable of handling the current amounts of runoff.

Reforestation has helped to deter landslides and rock slides, although more conclusive proof will be available when the trees have grown larger. Extensive re-growth and the return of life has occurred. The primary and secondary species provided enough soil nutrients to support a variety of trees and larger plants. Over time, flora and fauna have returned to the area, with bird-carried seeds adding to the biodiversity of the growing forests. Fruit trees have contributed to a greater diversity in the diets of favela residents. Water springs were regenerated, especially in the reforested areas in the West of the city, where fish and amphibians can now be found. A noticeable micro-climatic change occurred, with reforested areas providing shade and improving soil and air quality. A fall in temperatures was noted in testing and field observations.

The program's success can also be noted in the lives of the favelas residents. Since many resident were migrants from the countryside, planting and tending fruit trees restored a strong cultural connection to their agricultural past, while offering a more nutritious diet. The project revived an agricultural sector important to Rio's food supply. The use of paid community labor assisted favela residents in meeting their basic needs of food, clothing, shelter, and education. By providing jobs close to home, the project allowed for more leisure time, and did not force workers to spend a large portion of their earnings on transportation. The job skills gained from this experience offer participants future opportunities in other forestry positions. Ultimately, the project instilled a sense of community participation and raised political consciousness among the favela residents. Completion of the reforestation project in coordination with SMDS can, through individual and group empowerment, provide the momentum for other efforts to improve their communities.

As a result of its initial success, the Self-Help Reforestation Project has been established city-wide, growing from the original 25 targeted favelas. Public awareness of the project has promoted its further institutionalization in Rio and replication elsewhere in Brazil. Reducing risk and improving the quality of life for favela residents, this top-down program has successfully involved poor communities in a partnership which benefits both the city and the low-income settlement. This partnership is the root of the program's success both as a reforestation project and as a poverty alleviation project.

In terms of applying the Self-Help Reforestation Project to other locations, several factors must be applied. Municipal governments must be willing to make a strong commitment to such a program including providing or leveraging extended financial support. A commitment to reforestation requires access to adequate capital and planting research and technical site preparation. Average costs in the field, including labor, materials and transport for implementation and maintenance, were approximately US$ 4,900 per hectare. Since it has proven much cheaper and more feasible to hire local labor than to contract a private firm, while at the same time garnering local support for the project, efforts to undertake a similar program should include local paid labor as a vital component.

An essential element of replicating this project is the political will of municipal governments to work with low-income communities, particularly respecting local labor practices and land uses. When community leaders effectively operate with program advisors and local workers, residents' organizations can deliver the support of community members. The Self-Help Reforestation Project enforces these efforts by emphasizing reforestation as a means for local economic development and community empowerment. Community involvement and public education are therefore necessary for a successful program. Attempting to involve local residents, for example, through the use of local labor, may come against certain constraints related to labor laws. If this is the case, the government must be willing to alter labor legislation or find ways to get around it.

Several organizational structures are also key in efforts to transfer the project. The creation of a separate entity which is free of political ties and constraints, such as the SMDS, can greatly facilitate the rapid and efficient functioning of a program. Working with such an organization, an essential component to replication is the creation or support for strong, representative community groups. Without active and representative participation, reforestation efforts hold little chance of being respected by local communities, who can easily cut down the trees if they see no reason for their growth.

Urban Market Gardens: transforming unused land and poorly maintained drainage systems into vegetables in Accra, Ghana

Background and context

Ghana has a dual economy consisting of a traditional agricultural sector and a growing urban market sector. With decreased employment available in rural agriculture, Accra's growth has been stimulated over the last several decades by rural-urban migration. The city is experiencing a rapid annual population growth rate of approximately 5 per cent, and the population is projected to reach four million by the year 2010. The world economic recession of the 1970s and 1980s fueled this process by creating even more difficult economic conditions in rural areas and increasing the impoverishment of an already poor rural population.

Accra became a magnet for the economically active, including both local and foreign industrialists and manufactures, in addition to workers. In spite of this economic development, formal employment failed to keep up to the population growth. By the year 2000, the government predicts that the manufacturing sector in Accra will only provide one quarter of the population with employment. The urban center possesses neither the employment base, the infrastructure, nor the social services to support massive migration, exacerbating conditions of poverty.

The high population density has resulted in congestion, overcrowding, substandard housing, inadequate education and health facilities, poor sanitation and a generally degraded environment. Awareness of the nexus between urban poverty and the urban environmental in particular has become a priority concern in Accra. Many of the worst features of urban poverty are environmental in nature such as poor access to safe water, unsanitary conditions, smoky kitchens, contaminated food, uncollected solid waste and insect infestation.

Currently, about 48 per cent of the metropolitan population have income levels below the World Bank's absolute poverty threshold of US$ 307 per capita per annum. Accra's below or at minimum wage workers often need a secondary source of income, while the unemployed seek options in the informal economy. High food costs, resulting from low rural production and high transportation costs, also make it difficult to survive on a single income or maintain a nutritional balance. Surveys from the Ghana Living Standards Measurement Study show that, in 1987, urban households spent about 60 per cent of their income on food. Carbohydrates form a high percentage of food intake, as vegetables have traditionally been relatively expensive and considered a luxury item.

Another of Accra's critical problems is the management of waste water and drainage throughout the city. Industrial, commercial, and residential waste water discharges into open drains and flood channels. Responsibility for maintaining waste water and drainage lies with a number of local and metropolitan authorities. However, due to inadequate financial resources, weak management capacity, and the lack of well trained and motivated personnel, these agencies are able to provide only the barest minimum of sanitary needs to most of the territory under their jurisdiction. The physical infrastructure necessary for drainage is insufficient, poorly maintained and in some cases non-existent. The resulting health and environmental hazards are severe.

Project description

The Urban Market Gardens, a project of the Vegetable Growers Association of Accra, provide the city with vegetables and create supplementary income for low income workers by assisting them in transforming urban waste land into vegetable garden plots, which are watered with waste water they have filtered themselves. It is an innovative attempt to create secondary or even primary income, which simultaneously improves the urban environment by upgrading previously unattended land and adjacent drainage canals.

Originated as a response to the need to make extra income, vegetable production is largely undertaken by migrants from Burkina Faso and the Northern regions of Ghana. In 1987, in reaction to rampant evictions by landowners, the market gardeners came together to form the Vegetable Growers Association. At that time, the Association had 300 members. The number is currently estimated at 400 gardeners, and the Association represents all the market gardeners in the city. Since its formation, market gardening has become a legitimized practice and is receiving much more support from city officials.

Gardens are cultivated on small, inaccessible, unserviced, and vacant areas, averaging between 200 to 400 square meters. As a way to avoid maintenance responsibilities, property owners give free land use to gardeners, often without written terms of agreement. In exchange, gardeners occasionally offer them free produce. Gardens have also been set up on government properties. Much work must be done on the land prior to using it for gardening purposes. It must be terraced and stabilized against erosion, including carting in soil from other areas to build up eroded banks. Most of the work is done by a single gardener who personally finances the plot, as formal investment is absent. Gardeners rarely use family or hired labor.

For irrigation and other reasons, market gardeners attempt to find plots which are located adjacent to culverts, canals, and storm water channels. In order to utilize the waste water for irrigation purposes, they construct filtration gates. Once established, the filtration system provides a cleaner supply of water for household use as well. The gardeners also comb the waste stream to gather organic material to be composted for fertilizer. In addition, they gather discarded inorganic material and industrial matter such as auto parts, machines and plastic material to manufacture farming tools and create fencing material.

Once a plot and an irrigation source has been established, the actual gardening works begins. Crop selection makes the best use of site and location factors and resources in the area. Crops grown include those of European and local origin. Intercropping of small growing, quick maturing crops like lettuce and radish with larger, later maturing crops such as cabbage and cauliflower is common and offers high yields.

Market women who sell vegetables in the city are the main customers of the gardeners, followed by expatriate residents who visit the gardens. Gardeners also maintain client-salesman relationships with Accra's high-income residents, who they sell produce to at a higher price. The gardeners prefer to sell whole beds of vegetables instead of selling to individual buyers in small units. All produce are sold at the farm and no vegetable is stored. Marketing information is largely obtained verbally from fellow gardeners and from personal visits to stalls in town. Vegetable prices are not fixed, but are arrived at, after bargaining and haggling.

The gardeners have created important relationships at a variety of levels. They have been in contact with the Ministry of Agriculture, who have even set up seed supply shops throughout the city. Through savings accounts at the Agricultural Environment Bank, some gardeners have established important connections to the formal business sector, while gaining access to capital.

The Vegetable Growers Association has played an important role in promoting the credibility and viability of the market gardens. As mentioned, issues of land tenure and government fines were the impetus for the formation of the association. Because most gardeners did not apply for normal permission from the government to utilize unused city land, nor held written permission from private land owners, they were easily subject to eviction. The association were able to get the laws changed to allow the gardeners use of vacant land, and they have promoted a conspicuous absence of fines and taxes on the gardens. The association also networks with other cities agencies, for example with the extension service of the Ministry of Agriculture and the National Service Secretariat. Each year, at least four national service officers with agricultural backgrounds are sent to several market gardening sites to impart their knowledge and expertise, and to learn from the gardeners.

The Association also provides individual gardeners with technical support and counseling services. For instance, they offer farm spraying at minimal cost. In addition, they open an account with the Agricultural Development Bank, and use the funds from the account to support gardeners in purchasing seeds, tools and chemicals, and in rehabilitating farms in the event of destruction of beds due to flooding.

Obstacles

Initial obstacles to the Urban Market Gardeners came from land tenure and government restrictions. Local officials sporadically attempted to enforce land use regulations, although these random evictions rarely prevented gardeners from starting elsewhere. However, after the formation of the Vegetable Growers Association, they were able to minimize evictions, with the reasoning that drainage ditches and vacant land were being informally maintained by gardeners. As mentioned, they have been further able to promote a productive relationship with government officials.

Other obstacles came from the antagonistic relationship between gardeners and the Ministry of Health. Bacterial contamination of vegetables had led the government agency to prohibit gardening. Again, after pressure from the gardeners, the Ministry of Health looked for ways to work cooperatively with them. Instead of shutting down operations, they have assisted in upgrading irrigation water quality. Rather than fight an endless battle to prohibit and evict gardeners, government agencies have found it more productive to work with them to resolve the problematic aspects of urban marketing gardening.

Impact

Through simple vegetable production on marginal strips of land, gardeners are now able to provide Accra with 90 per cent of its vegetables. The average daily income of gardeners is three times higher than average daily wages in the formal economy. The Urban Market

Gardens have also provided greater economic opportunities for market women and expanded this previously small sector of Accra's economy.

The environmental impact of the Urban Market Gardens has been extensive. The poor state of Accra's storm and waste water drainage systems invites improvements by the market gardeners. Gardening and terracing, and the upgrading of banks and channels prevent erosion of canal walls and have saved the local government costly dredging and repair. The use of simple filter gates prevents flooding and improves water quality for more than just the gardeners. The recycling efforts of the gardeners of both organic and inorganic waste has cut down Accra's waste, albeit by only a small amount. Private landowners have also benefited by avoiding negligence fines for unused land.

The impact of the Urban Market Gardens, and in particular the efforts of the Vegetable Growers Association, has also been felt on the public policy arena; prohibitive and restrictive municipal regulations on land use, sanitation, marketing and water usage have been lifted. In addition, the Association has coordinated the transfer of technical support from government agencies to gardeners, improving their efficiency and output. The Association has continued to served as a policy advocacy group in municipal venues, offering a level of participation not known in Ghanaian politics.

Scaling up the impact

The appeal of the marketing gardening innovation lies in its simplicity and universal application; many of the urban poor in cities throughout the developing world are generally equipped with the agricultural skills to develop under-utilized land in their cities. Given the beneficial nature of gardening to a city, government officials should be willing to promote it. The increased access to low cost produce would no doubt gain the support of residents as well. The innovation requires only minimal and primarily private investment, and would thus not present any budget constraints.

The elements of Accra's innovation which have the greatest potential for application in other locations are those which integrate agricultural land-use with waste water and drainage maintenance. Though the mutually beneficial relationship that has developed between the Gardeners Association and the municipal government in Accra depends to some degree on the ineffectiveness of the agencies responsible for maintaining the drainage infrastructure, urban gardening can be compatible with well maintained sewage systems. Such usage would have a positive impact on the environment while reducing the costs of maintaining the infrastructure.

In order to adapt the program in another city, a number of pre-conditions are necessary. Marginal land, ideally located next to sources of waste water must be available, as well as a market for vegetables. If appropriate for the situation, partnerships should be created with the eventual vegetable sellers, such as the market women. The public sector must be willing to support this kind of land use, even to the extent of changing regulations and laws if necessary. This could for example require pro-active land use planning which incorporates small-scale, scattered site, urban agriculture into its drainage and water treatment systems.

If available land is primarily privately owned, gardeners must also be able to offer an incentive for allowing them to use the land or have a means to defend the use of private land. An association representing the workers must be formed to ensure that their interests are addressed. Finally, support is needed from a variety of public sectors, so that efforts are made to work cooperatively towards a common objective.

Lessons learned

Each of the innovations described represents a unique approach to an urban environmental problem. Each unfolded according to its own particular life cycle. There are, however, certain cross-cutting lessons regarding the origin of these innovations, their implementation, the obstacles overcome, the overall impact, and the potential for these to achieve scale. The lessons learned could be of great use to groups and individual seeking more creative and cost-effective solutions to urban problems.

Origins of the innovation

Where do innovative ideas come from? How do they get started? Four patterns emerge from the case studies:

1 The most fertile source of innovations are those cities and neighborhoods experiencing the most severe urban problems. The Zabbaleen innovation, for example, arose in large part because of a waste management emergency in Cairo. The city's trash collection needs had begun to outstrip the capacity of the Zabbaleen to fulfill them.

2 The most likely innovators are the grassroots leaders and the street-level bureaucrats who are closest to the problems on a daily basis. The Reforestation Project in Rio de Janeiro was initiated by the line workers at the Municipal Secretariat of Social Development in conjunction with the residents' associations of the first squatter settlements in which they worked. The Accra Market Gardens were created by an association of 300 vegetable growers in reaction to evictions by landowners.
3 Most innovations are driven by a charismatic leader or "product champion" who is tenaciously dedicated to implementing the new idea.
4 Successful innovations are derived from rigorous analysis of the current situation, past failures, and successful initiatives implemented elsewhere.

Implementation

How does innovation move from idea to implementation? Some of the cross-cutting patterns which emerged from our cases:

1 Successful implementation involves multi-sectoral partnership. The Zabbaleen initiative, for example, involved a collaboration among the Cairo Governorate, a private consultant, NGOs, and community groups. In Accra, the Urban Market Gardens Project is a collaboration among labor associations, landowners, local government, and private food distributors.
2 Sustainable innovation depends on direct community participation. In Rio de Janeiro, the local residents participated by designing the plans for their streets, gutters, and hillside orchards as well as physically creating them.
3 Innovators are distinguished by flexibility and openness to new ideas. Very few of the innovations followed a pre-planned evolution, but instead unfolded in unpredictable and sometimes surprising ways.
4 High impact innovations lead to integrated strategies. The Zabbaleen initiative grew from recycling/income generation into a comprehensive community development effort including: health care services, credit access, a mechanization program, women's literacy classes, infrastructure upgrading, etc.

Obstacles

By definition, innovation challenges the *status quo*. The cases show us that:

1 Innovation requires a fundamental change in habits, and the recognition that short-term risks and losses can result in long-term gains. In Rio de Janeiro, the Reforestation Project entailed considerable short-term work, both for the local residents and the government trainers. But it resulted in a new and sustainable grassroots/government partnership and a sense of self-sufficiency within the squatter settlement.
2 New practices involve reallocating resources and threaten established powers. In Cairo, the Zabbaleen scavengers were transformed from stigmatized social outcasts to key players in the environmental management of the city.

Impacts

In addition to the specific impact of each innovation studied, two overall trends are evident:

1 Innovation requires short-term, visible success in order to gain momentum for the long haul.
2 Collaborative partnerships are both a requisite for an outcome of the most powerful innovations. In Accra, the Urban Market Gardens Project resulted in a public/private/grassroots partnership with the capacity to continually generate solutions to Accra's economic and environmental problems.
3 In Cairo and Rio de Janeiro, local government and community groups have now established ongoing relationships for urban services delivery.

Scaling up

Many of the innovations in this paper are already being replicated in their own cities. From these experiences we can begin to hypothesize about the process through which urban innovation achieve scale:

1 The emergence of multiple leaders is critical to program replication: it "gives away" ownership of the idea and enables it to expand freely and be adapted to new contexts. Similarly, the first step in replicating

Rio's Reforestation Project has been the identification of established leaders - who understand and appreciate the innovation - in each of the squatter settlements of the city.

2   Two paths to creating impacts of scale from community-based practices are: horizontal transfer (peer-to-peer replication) and vertical transfer (scaling up into public policy); the Zabbaleen initiative moved from pilot project to large-scale initiative through incorporation into public policy (vertical transfer).

3   Transfer occurs along a continuum of integrity to the original initiative. In the Zabbaleen innovation, it is the basic idea which is distilled to its essence and diffused widely. In virtually all cases of replication, there needs to be some degrees of adaptation in order for the innovation to take root in its new setting.

4   The "import method" is more effective than the "export method". The replication of the Los Angeles Small Business Toxics Minimization Project in Rio de Janeiro State Government is being driven by the Rio de Janeiro State Government, because of their eagerness to find innovative toxics reductions strategies.

These case studies are small scale. The question is the upscale magnitude as a mega-city problem. The Mega-Cities Project as a transnational NGO continues to work on both these approaches.

8 The experience of regional environmental management in Bio Bio, Chile

Bolivar Ruiz-Adaros

Introduction

The Bio Bio region, located in Chile's central area, features a surface area of approximately 37,000 km^2 and a population of 1,732,000 inhabitants. It displays a rich variety of resources and natural beauty, but it also has huge contrasts. On the one hand, Bio Bio has plentiful natural resources, which support an important business sector related to their export, and a wide variety of landscapes ranging from practically unexplored beaches, abundant agricultural valleys and mountains slopes covered with natural forests, to a majestic mountain ridge. On the other hand, prevailing social and environmental data place this region among the poorest and most polluted in Chile, lacking facilities and featuring a low territorial integration with the rest of the country. The latter factor has brought about a significant decrease in living standards to an important portion of the national population (13 per cent), as well as a great impact on the national territory.

The natural element that defines and provides an identity to this region is the Bio Bio river basin, which drains 60 per cent of the regional surface and is an essential resource for the region's industrial, energy and urban development. In administrative terms, the region is one of the 13 units in which the country is divided. Its internal division features four provinces (Concepcion, Nuble, Bio Bio and Arauco), as well as 52 communes (*comunas*). Population distribution in the territory is quite uneven, with some of such communes having a high population density, while others are very sparsely occupied. This is partly due to the natural conditions, which create opportunities for better settlement in the coastal and central parts of the region, while poor weather conditions prevail in mountainous areas. Another reason for such an uneven territorial settlement is the fact that productive activities tend to concentrate in specific areas of the region, thus polarizing the occupation of the territory.

Out of the 52 communes comprised in the Bio Bio region, 15 of them have less than 10,000 inhabitants, 28 range between 10,000 and 50,000 people, five have from 50,000 to 100,000 inhabitants, while only four others extend beyond 100,000 people. Together, Concepcion and Talcahuano amount to 779,000 inhabitants, making for quite a dynamic economic urban center, while also acting as regional centers. However, this kind of uneven population distribution in the territory not only affects this region, but it also reflects a national characteristic, since 40 per cent of the Chilean population lives in the Metropolitan Region of Santiago, while, out of the remaining 60 per cent, 30 per cent is distributed between the Bio Bio and Valparaiso regions, leaving another 30 per cent of the population spread out in the 10 remaining regions of the country.

Such an unequal territorial settlement has created urban conglomerates, bringing about low living standards to their inhabitants, who must devote a great deal of time commuting long inter-urban distances, with a large number of them lacking access to recreation and amusement areas. Likewise, large cities concentrate so-called "hard spots" of poverty, the low level of development of which makes it difficult to put some of the government's social and assistance programs into practice.

The Bio Bio region concentrates a huge amount of the country's forestry surface, amounting to about 1,204,237 hectares of woodland in 1994. Besides comprising almost 50 per cent of the country's planted surface of pine trees, it also includes an abundant and rich coastal zone where 52 per cent of Chile's fishing activity takes place, corresponding to 3.6 per cent of fishing activity in the whole world. The region also contains steelworks, petrochemical and metal-mechanic industries, as well as shipyards and decreasing coal mining activities.

Some traditional agricultural activities can also be found in the region, being related to the farming industry and corresponding export products. The Bio Bio region has a range of over 400 types of exportable products, but nevertheless the greatest amount of exports corresponds to the forestry sector (logs, cellulose, timber and lumber wood, cardboard and paper) and to the fishing industry (fish meal, canned and frozen fish). The great variety of activities carried out in the region brings about a range of complex environmental problems, which must be managed through a joint effort by public, private and community organizations.

Environmental problems in the Bio Bio region

Water and air pollution, soil pollution and erosion, and the loss of bio-diversity may be considered among the major environmental problems that affect living standards and economic production in the Bio Bio region.

Water pollution

Water pollution comprises physical, chemical and organic pollution at surface water level (rivers, lakes, ponds and sea) because of industrial and residential sewer system discharges carried out without any kind of treatment, since there are no water treatment plants in the region, neither for domestic nor for industrial purposes. Other forms of pollution refer to the eutrophy of ponds, bays and estuaries produced by an excess of organic matter in masses of water, and to the underground water pollution caused by the use of pesticides and fertilizers in agricultural and forestry activities, besides the infiltration of filtered liquids in improper sumps and dumping grounds.

Air pollution

Air pollution in Bio Bio is mainly produced by the development of industrial activities using technology which does not include suitable filter systems, and that intensively uses raw materials and supplies likely to produce great volumes of volatile material. Besides which, there are some types of businesses, such as fishing companies, that produce a kind of foul-smelling pollution which causes serious discomfort to the whole population.

Atmospheric pollution increases with the intensive use of fuels such as coal or firewood both at industrial and domestic level. This kind of fuel easily generates particles and contributes to the external build-up of other airborne products such as sawdust, ash and clinker.

Soil pollution and erosion

Physical, chemical and organic pollution is caused in Bio Bio by the improper collection, treatment and final disposal of urban solid residues, both industrial and domestic, as well as by the inadequate spillway system for excrement disposal and the improper use of such systems as receivers of industrial liquid waste.

The use of fertilizers and pesticides greatly contributes to the problem of soil pollution in the agricultural and industrial sectors. Ground erosion, mainly in the coastal mountain ridge, in the Andes Mountain slope areas and the central valley, happens because of poor agricultural and forestry management, as well as due to the following factors: the utilization of some farming procedures on slopes in which inappropriate techniques are used; intensive tree felling carried out by some forest companies (most of which belong to foreign corporations); use of fire to clear areas; and firewood removal. As a result, the soil has become exhausted, and large areas of the region have become useless. The extent of such loss has not been quantified yet.

Loss of bio-diversity

Another significant problem is the loss of bio-diversity of existing flora and fauna as a consequence of the replacement of the natural forest by pine and eucalyptus monocultures. Loss of bio-diversity in fishing resources is due to overfishing beyond reproduction rates, resource overexploitation and the degradation of the areas in which some species inhabit. Pollution and bio-gathering of pollutants in the tropic chains are also factors which should be mentioned.

The abovementioned problems, which were briefly presented, are highly complex both in regard to the reasons that produce them and to their effects on the population's living standards and on the environment's condition. Such problems take different forms depending on the area where they develop. Among the region's most critical situations, Talcahuano and Coronel should be stressed, as they are densely populated and highly industrialized urban areas that have developed on rather small surfaces, where various kinds of activities coexist in an irregular way. Likewise, the abovementioned communes are characterized by high poverty levels affecting more than 40 per cent of the whole population, living in a highly polluted environment.

Other problematic environmental situations are those caused by the cellulose industries located in very low populated areas. Those companies show little concern about the area where they are established, and they cause serious damage to the environment, polluting the air, soil and especially water, because they pour their waste products - without being properly treated - into the sea and rivers. Exploitation and degradation of natural resources leads to other critical problems often less understood, but which potentially lead to loss of potential and impoverishment in the medium- or long-term.

Historically, environmental problems worsen due to the fact that the law in force was not updated to cover such situations. All the abovementioned problems also seemed to increase due to facts such as the limited funds assigned by the government to environmental preservation, the short-term viewpoint of productive organizations that try to increase their earnings and competitiveness by spoiling the environment, and the limited education and civic action of the community, which often view these issues as being beyond their control.

Another outstanding factor to be considered is the still centralized and concentrated nature of Chilean governmental policies, in which the different regions have reduced powers of self-determination to solve their own problems and manage their own development. This is due to an irritating tradition of centralism, with the central government being responsible for managing public funds, but it is aggravated further by the limited scope for action, decision-making and management by regional officials, who are used to receiving guidelines from the central level. As a result, those who have a better and clearer picture of the problems are not the ones in charge of solving them.

Although this situation has happened throughout the history of the country, the arrival of democracy in 1990 brought about some changes in the way such problems are tackled: more specific decisions have been made such as the transfer of funds to the regions, a constant transfer of powers and functions and the settling of regional offices for some traditional centralized services, in which regional officials have limited decision-making and management capacity.

The governmental strategy to manage environmental problems

The recent governmental strategy aimed at managing environmental problems in Chile, and in the Bio Bio region, is based on the formulation of a legal framework, the creation of a responsible environmental agency, and the enforcement of various environmental management instruments and powers that have been recently introduced by law. Both the legal framework and the agency created - the National Committee for the Environment (*Comision Nacional del Medio Ambiente-CONAMA*) - have been established to coordinate a new management system which, however, takes into account the responsibilities and powers of traditional organizations and agencies.

The new national environmental law was put into force in 1994, aiming to materialize an existing constitutional precept according to which every Chilean has the right to live in a pollution-free environment. It must be noted that, before the enacting of this law, there were about 1,200 legal texts regarding environmental matters in force, some of them dating to 1916. In addition, about 14 governmental ministries and 30 public agencies or offices had specific environmental responsibilities. Nevertheless, and in spite of that, the environment as a whole was experiencing an increasing degradation process.

One of the factors that caused the abovementioned situation was the lack of a governmental agency responsible for coordinating both the ministries and public agencies that have some authority on environmental matters, the duties of which were spread out across a huge variety of different organizations operating in uncoordinated ways, and often featuring parallel and ambiguous functions and responsibilities. Furthermore, there was a lack of an all-encompassing national policy aimed at providing coherence to the scattered environmental legislation then in force in Chile.

The 1994 environmental law confronted such problems by creating an agency responsible for environmental preservation and by articulating the different instruments available for environmental management. Among the new instruments created by this law, the definition of environmental quality regulations, the requirement of decontamination plans, and the assessment of environmental impact in new projects deserve special mention.

Given the lack of regulations and the need to take immediate action relating to environmental matters, a decision was made to voluntarily apply some of such instruments, for example, the evaluation system for environmental impact, to which about 160 projects have been subjected by now. This has made it possible to test the coordination capacity of some public agencies, as well as to detect further requirements for refining the system. Likewise, and with the purpose of developing practical experiences and solving real problems, the requirement of an "environmental recovery plan" was introduced, which was applied to the abovementioned city of Talcahuano in the Bio Bio region.

It should be stressed that an "environmental recovery plan" is not the same as the abovementioned "decontamination plan"; a decision was made to formulate the recovery plan in Talcahuano because the application of the latter requires that the emission regulations must be first exceeded, which in Talcahuano has been impossible to verify since no measurements have ever been performed.

The Environmental Recovery Plan for Talcahuano

The Environmental Recovery Plan for Talcahuano (PRAT) was an experience of regional urban and environmental management in which the main component was the participation of all social actors at all levels, ranging from the diagnosis of the existing problems to the proposition of solutions. It lasted for 17 months, from August 1994 to December 1995, and US$ 300,000 were invested in it. The plan was financed both by CONAMA and the city's private industrial sector.

The successful outcome of PRAT's different stages rendered possible the elaboration of an updated diagnosis of the environmental and urban realities of the Talcahuano commune and the formulation of proposals comprising 30 project profiles as well as 70 basic actions for the "Action Plan for an Environmental Recovery in Talcahuano". The environmental framework behind the project included all aspects of territorial organization, natural resources, pollution, health, education and participation.

The actors who were directly involved in this process also took part in the formulation of proposed solutions, both those affected by environmental problems and those who are responsible for them. The result of such interaction was a realistic and participative management process, which makes it more likely that the plan will be implemented. PRAT is currently at the execution stage, and some specific actions have already been undertaken in the city aiming at improving their inhabitants' living standards.

PRAT was the first such experience to be developed in Chile in which the whole population of the city (250,000 people) was considered to be part of this joint action. It was financed both by public and private funds, and more than 100 professionals took part in it at different levels. This certainly means that it is a most successful experience of urban and environmental management, which should be considered for other cities or similar administrative areas.

Background

The Environmental Recovery Plan for Talcahuano was conducted by CONAMA, and was widely supported by Talcahuanos's Municipality, as well as by other public organizations, public and private enterprises, and some community organizations. As was mentioned above, PRAT became a pilot project because of the regional community's concern with the process of environmental degradation that affected Talcahuano at the time. At a meeting held in May 1994, CONAMA discussed with businessmen and university professors the idea of declaring Talcahuano a "saturated area" in the context of pollution, a new concept introduced by the Chilean environmental legislation.

Shortly before, in March 1994, a huge fire had affected San Vicente Bay, a harbor located in Talcahuano City, which is used as an oil, industrial and fishing harbor as well as for the export of forestry products. One operator was killed and the businesses in the area were physically and economically damaged; the local population was overwhelmed by panic, due to the possibility that some fuel tanks in the area could have burst. As mentioned above, the area is overcrowded and polluted by the industrial activities performed there. The combination of factors such as community accusations of health decline, disputes due to the bad smell arising from the fishing companies' activities, gas effluvium from a company nearby that produces domestic gas from coal, as well as other problems that arose from the inadequate use of land by some small and medium sized companies, created a most negative background of environmental and social conflicts.

These factors encouraged CONAMA to undertake a pilot plan in order to start a process of environmental recovery in the city. The strategy of such a pilot plan was to make a diagnosis, formulate proposals and elaborate an action plan in which all agencies and organizations would have a specific task to fulfill.

Once this action plan was submitted and approved by all PRAT partners, the proposal development stage concluded. The Technical Office as well as the Coordination Committee completed their task at this point, and a new phase started, in which every entity involved is responsible for fulfilling those tasks which they were attributed. It is important to stress that each of the entities or individuals involved joined and adhered to the agreed program on a voluntary basis.

The legal framework

The legal framework supporting the plan was based on the mandate given by the 1994 environmental law to CONAMA to be responsible for the formulation and implementation of the national environmental policy. This law also clearly stated that CONAMA is responsible for enforcing measures aimed at protecting the environment, decreasing pollution and preserving natural resources. Besides which, the Presidency's General Secretariat Office is committed to confronting the pollution problem in Talcahuano as a top priority issue. Such commitment was confirmed in the PRAT's opening session in August 1994, and in the closing ceremony in December 1995. Both events were attended by the wider community.

Institutional design

The plan was organized around two main bodies, a Coordination Committee formed by local authorities as well as representatives of the city's main organizations involved with environmental matters, and a Technical Secretariat responsible for doing research as well as coordinating the activities of the different commissions that developed the proposals sustaining the action plan.

The Coordination Committee was headed by the Regional Intendant, who is Bio Bio region's principal authority, its vice-president was the City Mayor and its Executive Secretary was the Regional Director of CONAMA. The other members were: the Maritime Governor, and representatives from the National Health Service, petrochemical and the steelworks sectors, industrial fishing companies, and the harbor administration. It also had the participation by representatives from the *Union Comunal de Juntas de Vecinos* (representing residents' associations); the *Accion Ciudadana por el Medio Ambiente* (a citizens' organization concerned about the environment); ESSBIO (the company that supplies drinkable water and manages the sewer system); CUT (the official workers organization in Chile); Huachipato's Trade Union (the largest steelworkers' union in Talcahuano), and finally, the Production and Trade Chamber.

The Technical Secretariat comprised an inter-disciplinary group of professionals, including engineers, physicians, biologists, geographers, architects, sociologists and teachers. The work was organized in themes: territorial arrangement, natural resources management, public health, environmental pollution management, environmental education and citizens' participation.

Objectives

As was mentioned above, PRAT's general objective was to define an action plan aimed at reverting the environmental degradation process in the city of Talcahuano as well as to establish the foundations for its sustainable development, in order to improve living standards and to protect the population's health, to recover natural resources and to make the community's productive system compatible with the other activities common to the urban environment. The program's specific objectives were the following: to detail an updated diagnosis of the city's environmental situation; identify the best solutions to the problems detected; propose an action plan linking the suggested solutions; define a mechanism to ensure the execution of the action plan; coordinate activities of environmental education and training with different sections of the community; and coordinate the plan's promotional activities in the community.

Methodology

Diagnosis The diagnosis was carried out by gathering bibliographic information and by the undertaking of studies related to specific topics, with the purpose of detecting "problematic situations". Each situation was introduced according to a theme (territorial organization, natural resources, pollution, health, education and participation) and described within the scope in which the problem was best explained (natural/ built/physical, cultural/social, normative/political, productive/economic, institutional/administrative, and technological/scientific). The diagnosis was subjected to discussion by a wide range of groups, and the problems were prioritized by them.

Proposals Working committees from different sectors were appointed for each of the abovementioned themes, having developed the proposals in workshop sessions. The methodology used in this stage was the so-called "objectives aimed planning". Firstly, the main problem is defined, and then it is "broken down" in a kind of "tree" of interlinked problems, thereafter, the proposed solutions are pursued by building "matrixes for project planning" by means of which the whole proposal is formulated. The final proposal was subjected to discussion by the various groups mentioned above during an open seminar in order to evaluate it and to reach consensus.

Accomplishment The proposal was then incorporated into a joint action plan checked by a group of technicians from different backgrounds. The plan was approved by the people and organizations taking part in PRAT as well as by most important local authorities. A financing program was submitted to and accepted by the World Bank, by means of which the projects included in the action plan under CONAMA's responsibility were initiated.

145

Results of the action plan

The action plan is quite an extended document comprising specific actions and projects. Concerning the matter of territorial organization, the plan contained a land use proposal to be included in both the city's and in the inter-cities' master plan. Four macro-proposals were presented comprising 19 project files, namely: projects to empower the city's natural and cultural assets; projects to strengthen the roles of the harbor, residential and industrial sectors in the city; projects to improve urban management; and projects to improve transportation management.

Concerning the management of natural resources, a proposal was made about the use of the maritime and continental territory to allow for the preservation and management of coastal, salt marsh and ground ecosystems, and thus to keep the landscape-type value of the natural systems in the commune.

With regard to the issue of health, a general proposal was made that the management of the agencies involved in environmental matters shall be reinforced. In relation to pollution management, some actions were proposed to the chemical-petrochemical sectors, as well as to metal-mechanic and steelworks, industrial fishing companies, harbors, shipyards, and other industries of various activities in the city, in order to improve emission levels and data, the suitable process management aimed at decreasing the pollutants in their sources, and the final disposal of liquids and solids, along with the controls on gas emission and noise.

Regarding environmental education and participation, a proposal about environmental education was submitted, both at a formal and an informal level. At the formal level, it suggests the formulation of a pilot plan for interdisciplinary environmental studies at elementary schools covering theoretical subjects as well as special practical activities. In relation to the non-formal level, it suggests an environmental education program for both community organizations and those pertaining to a guild or trade. There was also a proposal aimed at empowering the community to participate in environmental matters by means of a formal organization that was to be technically supported.

In the action plan, CONAMA assumed a commitment to several specific matters, namely: to support the urban and environmental development of small and medium sized companies; encourage the creation of regulations concerning air and water issues, generating the necessary information through suitable monitoring; encourage projects for the suitable management of the existing natural resources in the city; support management systems for environmental information so that speedy and accurate actions can be taken; conduct environmental education programs; and support the generation of opportunities for citizens' participation in environmental matters.

In addition to the abovementioned attributions, CONAMA considered that it was necessary to coordinate the fulfillment by other agencies and organizations of the commitments that had been assumed in executing the plan.

Supporting small companies

To support the urban and environmental development of small and medium sized companies, a cooperation agreement was established between CONAMA and CORFO, so that the latter could open a financing line for projects related to environmental matters. This agreement is now in effect, and businesses that want to pursue technological improvement, information generation and investment in environmental matters may apply for subsidies amounting up to 80 per cent of the whole project. CONAMA has held several meetings to spread the availability of CORFO funds and thus encourage the small and medium sized businesses to invest in environmental improvement.

Since many such companies are not even aware of what the environmental problems are, CONAMA has supported the development of the FONSIP/PYMES project, to be carried out by INTEC Chile, to study the most important characteristics of the Talcahuano commune. This project, to be developed over a three-year-period, aims at diagnosing, training and managing urban-environmental solutions in industrial activities in the country's three metropolitan areas.

To complement the information produced by INTEC, CONAMA's Regional Head has submitted a project to CORFO to make a speedy diagnosis about some topics that have not been considered in the FONSIP/PYMES project, and that also provoke negative environmental impacts in the city. The purpose of this project is to generate background information which will make it possible to present future projects about environmental improvement by making use of funds supplied by CORFO. The topics to be discussed are gas stations, sandblasting and industrial painting factories, and part of the city's public transportation may also be included in the near future.

Regulating air and water

Another main concern was to encourage the creation of regulations of air and water quality, generating the necessary information through suitable monitoring. With regard to air pollution, some regulations covering emissions require the imposition of penalties for those activities which exceed approved levels. Such regulations have been developed in relation to measured concentrations and accepted limits that are not considered harmful for human health.

Of course, in order to make proposals for regulations it is essential to have proper measurements of industrial emissions. This encouraged CONAMA to request the Swiss International Cooperation to finance the "Preliminary Monitoring and Shaping of Air Quality in Talcahuano" project. The purpose of the project is to obtain information about the environmental situation in the area. By doing so, it will be likely to enable the formulation of plans for minimizing emission, especially in critical cases, for the targets to be reached, as well as the degree, priority and urgency of such decreases. Likewise, local institutions will be trained in environmental management, evaluation, emissions control, control of technologies, etc.

This project comprises three steps, with the final step consisting of the establishment of a monitoring network, including sensors that will provide information about emitted compounds and their concentrations, with the support of foreign experts (University of Stockholm and INDIC AB) as well as Chilean ones (Instrumental Analysis Department, University of Concepcion, EULA). The total cost of the project exceeds $ 246,000,000 (Chilean currency; US$1=$410 Chilean Pesos). Financing for Phase 1 of the project has been obtained, amounting $ 10,250,000, the final report of which has already been finished. Parallel to this, the second phase of the project, amounting $ 12,300,000 has been approved and started.

Concerning hydric resources, CONAMA's Regional Office has suggested the formulation of a "Monitoring Program for Concepcion and San Vicente Bays", the cost of which amounts $ 16,000,000. The purpose of this program is to collect environmental background information to create a preliminary framework for "Regulations for Environmental Quality" with respect to sea water in Concepcion and San Vicente Bays. A working plan submitted by one of the applying institutions to the development of the project has been approved.

Managing natural resources

There is a desire to encourage projects for the suitable management of the existing natural resources in the city. There are many important natural resources in the Talcahuano commune regarding biodiversity, such as ponds or lagoons, salt marshes, estuaries, humid soils and peninsulas. Among the measures that have been considered, the formulation of a management plan for the ecosystem of the lagoon called "Laguna Price" deserves mention. The purpose of this project is to elaborate a plan to guarantee the suitable management of the abundant fauna, especially birds, existing in this area. The cost of the project amounts $ 2,900,000 and the research will last about six months.

Another proposed action refers to the matter of humid soils in the region, many of which are present in Talcahuano commune. A project applying to regional funds has been submitted: "Diagnosis of the preservation status of humid soils in the Bio Bio region and a proposal for their sustainable development". Such a project, with an investment of $37,000,000, proposes a detailed study of the existing humid soils in Talcahuano. Comprised within the scope of PRAT, the discussion will focus on the criteria to be considered in the formulation of proposals for the use of these areas.

Finally, another proposed action related to PRAT should be mentioned: the "Design of the management plan for the nature sanctuary at Hualpen's Peninsula", which is a beautiful scenic area within the city's limits. A budget of $ 27,000,000 has been invested in the project, aimed at developing a management plan for this 20-year-old sanctuary. The project applied to regional funds in 1997.

Environmental information

To support management systems for environmental information so that speedy and accurate actions can be taken, two proposals were made, one of them with the objective of creating a regional territorial information system. This idea was adopted by PADERE (Regional Planning Office) and is being developed. The second idea consists of developing an environmental information system at the commune level.

In relation to the latter proposal, PRAT hired GEOSIG Company to develop an environmental information system for the city of Talcahuano. The purpose of this system is to optimize the handling and management of information on environmental matters in the commune, thus increasing the efficiency of environmentally-oriented actions especially by reducing the length of the decision-making period. The cost of the project amounted to $ 3,700,000.

Environmental education

An environmental education program for Talcahuano community began in 1996. The purpose of this program is to support citizens taking part in the local environmental management by means of the creation of various community groups involved with environmental matters. Young students and women are given first priority, since they exercise a multiplying effect within their families. A monitors' training course has been developed, and the choice of elected people has already occurred.

There was also a proposal to include environmental education as part of the elementary school programs at the commune level. All the elementary educational institutions in Talcahuano should have at least a copy of the text of the plans in force so that environmental education can be studied at a classroom level. Contacts with the corresponding authorities have been started.

Community action

A working plan for social organizations was proposed to support the linking of environmental with the other social organizations existing in the city. Contacts with these organizations have started, while, with the support of European environmental institutions, workshops, seminars and meetings are being held and will continue, thus creating a working plan with the social organizations performing on a joint basis.

Other actions

Other actions supported by CONAMA relating to Talcahuano's Environmental Recovery Plan include the feasibility research on the management of liquid industrial waste in Talcahuano; the monitoring and preliminary draft legislation about air quality in the city (Phases II & III); an environmental awareness program for children and teenagers in extremely poor areas; a discussion about the efficiency of air quality regulation; the building of a simple computer database aimed at the identification of productive processes in different locations of the Bio Bio region; and the identification of the characteristics of the historical and cultural heritage to create a baseline for the Bio Bio region's national monuments.

Other important actions include the formulation of a management plan for the abovementioned nature sanctuary at Hualpen, the acquisition of equipment for automatic air quality sampling, an analysis of the variables that generate environmental problems in the region, and a feasibility study of the disposal of industrial solid waste.

Environmental diagnosis for the coastal area of the communes of Coronel and Lota

The communes of Coronel and Lota are two urban areas that resulted from the coal mining activity, which is in decline. These focal points concentrate, as in Talcahuano, an important percentage of those people living in poor conditions.

The research done consisted of an environmental diagnosis based on the gathering of background information following very exact methods, thus obtaining sound conclusions. Once this information has been gathered, an environmental strategy will be formulated, once again considering popular participation as its more important aspect.

A critical analysis of the experience of regional environmental management

The events described above have been, in general, evaluated positively by the main social actors in Talcahuano, as well as by the authorities. There has certainly been a significant change. Companies, or at least most of them, have started to assume the responsibility for the environmental costs brought about by their activities. Public authorities have started to enforce the legal authority they have been granted in a more efficient and timely way, while the community, even though it seems sometimes to be somewhat indifferent, has progressively become conscious about the fact that, if they do not take part in their city's environmental management, they will not be able to improve their living conditions.

In fact, there has been a great improvement in most parts of the areas considered to be problematic. There are some new projects and actions, both public and private, that follow the directions suggested in PRAT's proposals. Nevertheless, and in spite of the above, some sectors have opposed this innovation. Some companies are still reluctant to make investments in order to undertake decontamination, there are still some coordination problems to be tackled, and a certain suspicious attitude by some the authorities themselves can be detected in some cases. The traditional attitude of the Chilean public servant tends to work within a specifically focused area and/or matter, as they have not been trained to work in a more integrated way, or on an inter-disciplinary basis. This, undoubtedly, has been an important disadvantage in this kind of local management action.

On the other hand, the administrative decentralization process has not become part of certain sectors and services of the Chilean governmental administration itself, which means that in some cases prompt decisions need to be taken to make them effective, but this depends more on the central level rather than on the local agents.

Specific results

Linking the environmental dimension to the broader discussion of the region's development was of utmost importance. At an enterprise level, there was, to a large extent, an acknowledgment of the existing serious environmental problems that have to be confronted. The project's extraordinary mobilization capacity should be stressed, as it has created a "meeting scenario", a forum that did not exist before, both in the stages of diagnosis and proposal. Normally, there is limited communication between the public administration, the productive-private sector and the wider social sector, but when discussing this project an excellent opportunity arose to have these different actors communicate, interact and definitely make real contributions aiming at solving a problem that affected all of them.

This is perhaps the most valuable outcome of this project: to have all the community joined together in order to find solutions - in the mid-term - for a long lasting problem, in contrast to the ineffective and non-strategic actions that had been taken over more than eighty years. Given the context of a wider anarchic and uncontrolled urbanization and industrialization, the past lack of policy resulted in Talcahuano's commune present pollution and degradation process.

Other important results were: the incorporation of environmental criteria into traditional mechanisms of land use and territorial management; the undertaking of projects and actions aiming to obtain information about regulation and management of natural resources; the participation of community organizations in all stages of the process; the creation of agencies which can provide the community with information concerning the current actions and proposals aimed at achieving environmental solutions for the city; and the involvement of a large number of teachers in non-formal environmental education programs. It is also important to stress the regional entrepreneurs' positive attitude when it came to sit for open discussions about environmental topics.

Among the many factors which have put obstacles to these new experiences of environmental administration, the tradition of centralism, the lack of political will shown by some authorities and the low number of technical environmental regulations and standards deserve to be mentioned, as they make supervision difficult. Other difficulties have also rendered the enforcement of the proposals difficult. PRAT was basically an operation-based concept, without specific legal attributions. It worked well in the stages of diagnosis and proposal, but the structure of the plan does not seem to be so effective as it would be required to bring these into practice. There are conflicting legal provisions concerning the distribution of attributions for the provision of services. More efforts have to be made to guarantee the operational coordination of public agencies, and also to minimize costs.

Moreover, incentives should be given to economic activities likely to attract new investments, so that they take environmental criteria into account. Most of the region's export products (almost 50 per cent) are bound to the Asia-Pacific area and have a low aggregate value. Since environmental requirements in those countries with respect to our regional products are low, companies do not feel the need improve quality in order to compete for the market.

Conclusions

The PRAT experience has not yet finished. Its main proposals have already acted as "raw material" to generate other proposals and a great amount of environmental recovery projects in Talcahuano city itself. A synergy has been created, which is undoubtedly a positive factor regarding the effort made by the community's various agents in search of the city's environmental recovery. We believe that the project has been a valuable tool for environmental management which has led to very positive specific actions.

The community believed in, and still trusts PRAT's proposals. That is why it keeps on taking part in different actions and activities comprised within the program. This has led representatives of the local business sector to consider regular attendance in various technical working groups in which CONAMA has invited them to participate. On other hand, the project's positive outcome has led other authorities in other cities of the region to try to reproduce it, with CONAMA's support. Some harbors in the region, such as Coronel, are the most interested parties in introducing

PRAT's methodologies and actions in their corresponding territories, to prevent the same environmental degradation that had affected Talcahuano. Undoubtedly, this is also a positive development.

Due to the above, both the national CONAMA as well as the Bio Bio region's CONAMA have made a decision to promote this new type of joint environmental management project in Chile, in which the whole community of the city takes part. Some economic resources have been already been committed, although they are not sufficient due to the magnitude of the task.

We are sure that the experience will continue, so that, eventually, Talcahuano, which happened to be one of the most picturesque harbors in Chile, will become again an attractive place not only for those who desire a decent and improving work environment, but also respectable and healthy living conditions.

9 Sustainable cities and local governance: lessons from a global UN Program

Jochen Eigen

Introduction

A sustainable city is a city where achievements in social, economic, and physical development are made to last. A sustainable city has a lasting supply of the natural resources on which its development depends using them only at a level of sustainable yield. It maintains a lasting security from those environmental hazards which may threaten development achievements, allowing only for acceptable risk. If sustainable cities are fundamental to social and economic development, broad-based local governance is key to sustainable cities.

The Sustainable Cities Program (SCP) is a practical response to the universal search for sustainable development. It is a United Nations (UNCHS/UNEP) joint facility to package and apply specialized know-how in urban environmental management. Being driven by local needs and opportunities, the SCP operates at local, national, regional and global levels. It is a vehicle for inter-agency cooperation, and it supports a learning-based process to advance collective know-how among the partner cities.

Sustainable cities and socio-economic development

It is now widely recognized that cities make an important contribution to social and economic development at national and local levels. They are important engines of economic growth, absorb two-thirds of population growth in developing countries and offer significant economies of scale in the provision of jobs, housing and services. Cities are important centers of productivity and social advancement.

However, the full realization of the cities' potential contribution to development is often obstructed by severe environmental degradation in, and around, the rapidly growing urban centers. For example, the development on land prone to environmental hazards such as flooding not only wastes resources but leads to further environmental degradation. This impacts most cruelly on the urban poor. Environmental degradation threatens economic efficiency in the use of scarce development resources, social equity in the distribution of development benefits and costs and sustainability of hard-won development achievements.

Broad-based local governance: the key to sustainable cities

The cities of the world, from Katowice in Poland to Dar-es-Salaam in Tanzania, and from Concepcion in Chile to Shenyang in China, are very different in their environmental setting, the type and level of their development, and the set-up and capacity of their administration. But most cities have in common enormous environmental problems and a strong commitment to resolve them. Cities also have increasingly in common a firm understanding that solutions to their environmental problems, to be effective and sustainable, cannot depend upon external or central government support but must rely upon local technical and financial resources. Mobilization and proper application of these resources from the local public, private, and community sectors requires new approaches to governance and urban management.

Stronger local governance through stakeholder involvement is the key to the effective creation of sustainable cities, as it leads to better conditions of environmental decision-making, better implementation of environmental strategies, enhanced administrative/managerial capacities, and better environmental information and technical expertise on available technical and financial resources.

For Africa, such approaches have been discussed and principles have been adopted by more than 20 governments in the Dakar Declaration (June 1995). Representatives from Asia, Latin American, and European cities met in Madras in February 1996 to review their own experiences. In June 1996, as part of the Istanbul "City Summit", the consolidated conclusions were presented by the cities to the international development community during a special meeting on "Implementing the Urban Environment Agenda".

The Sustainable Cities Program (SCP): a practical response to the universal search for sustainable development

Both the urban environment and local governance have been accorded unparalleled attention in the recent international debate on development. UNCED, the 1992 "Earth Summit", will be remembered as the conference in which the world acknowledged the importance of the environment for social and economic development. This was articulated in Agenda 21, which emphasizes cross-sectoral coordination, decentralization of decision-making, and broad-making participatory approaches to development management. UNCED also recognized the potential of the SCP as a vehicle of implementing Agenda 21 at the city level, and recommended strengthening its role in this regard. Habitat II, the 1996 "City Summit", took this point further in a global agenda for cooperation that acknowledges the direct contribution that sustainable cities can make for social and economic development.

In the urban environment, the mandates of Habitat and UNEP coincide, and their scientific, technical, and financial resources are uniquely complementary. In the early 1980s, the two United Nations agencies were instructed by their governing bodies to jointly prepare environmental guidelines for settlements planning and management; then, in the early 1990s, to launch the Sustainable Cities Program to put the concepts and approaches of the guidelines into practice; and more recently, in 1995, Habitat's Human Settlements Commission and UNEP's Governing Council instructed to transform the SCP into a truly joint facility for implementing Agenda 21 at the city-level.

In this context, the SCP is a UN facility to package and apply specialized know-how in urban environmental management. It provides municipal authorities and their partners in the public, private and community sectors with an improved environmental planning and management capacity. In the first four years since its launch, the SCP grew from a modest US\$ 100,000 per year initiative to a US\$ 20 million global program mobilizing support for many sources including UNCHS, UNEP, UNDP, World Bank, WHO, ILO, the Netherlands, Denmark, Canada, Italy and France. The SCP is undertaking demonstration projects in some 20 cities worldwide. These cities demonstrations result in the formulation of Local Agenda 21 which include environmental management strategies, action plans, and priority technical cooperation and capital investment projects for the cities concerned.

SCP activities, at various stages in the project cycle, are currently underway in cities such as Accra, Asuncion, Concepcion, Dar-es-Salaam, Dakar, Guayaquil, Ibadan, Ismailia, Katowice, Madras, Shenyang, Tunis, and Wuhan. Some cities are about to graduate from city level demonstrations and advance to national replications (for example, the Dar-es-Salaam demonstration is about to be replicated in Tanzania's eight secondary cities); some cities have just completed the preparatory phase and are ready to start fully-fledged demonstrations activities (for example, Madras and Tunis); yet other cities have recently requested participation in the SCP (for example, Lusaka and Maputo).

The SCP is driven by local needs and opportunities. It is pre-eminently a locally-focused program, in which national, regional, and global support is built up from activities and experiences at city level. The SCP provides a framework for linking local actions and innovations to activities at the national, regional, and global levels, through which the lessons learned in individual city experiences can be shared, analyzed, generalized, and discussed widely. This operational link serves to make global strategies more responsive to local needs and opportunities and, conversely, helps to implement global strategies and agreements at the local level. As a global program, the SCP promotes the sharing of know-how among cities in different regions of the world. As an inter-agency effort, SCP mobilizes, packages and applies technical and financial resources from diverse sources.

Operational levels

The SCP has four operational levels, namely: the city level; the national level; the regional level; and the global level. The primary focus of the SCP is at the city level where, in its initial four years, the program applied more than 95 per cent of its resources. The SCP provides a framework in which lessons learned in individual city demonstrations can help shape activities at the national, regional, and global levels.

At the city level, the SCP brings together all the skateholders whose cooperation is required to clarify environmental issues, agree on joint strategies and coordinate action plans; to implement technical support and capital investment programs, and to institutionalize a continuing environmental planning and management routine. SCP activities are conducted by municipal governments and their partners in the public, private and community sectors with technical assistance as necessary. The aim is to incorporate environmental management into urban development decision-making and to strengthen local environmental capacity in city demonstrations projects. SCP operational activities respond to local issues and needs as perceived and expressed by local governments, the private sectors and communities. City demonstrations are the focus of the work (lasting for two-three years, and costing US$ 1-2 million each), which also aims to help local practitioners and decision-makers identify priority issues, involve stakeholders, develop strategies, agree on action plans, and implement priority projects and institutionalize process.

At the country level, SCP activities are conducted by national governments and their public and private partners supported by local and international consultants. These activities are an essential complement to city-levels demonstrations. The aim is to address national issues arising from city-level experience and to promote replication strategies for cities throughout the country. Tanzania and Chile have reached this stage in the project cycle, with Egypt and Nigeria soon to follow. In the near future, more SCP demonstrations will advance from city to country-level activities. Global level support is being expanded accordingly to take full advantage of opportunities for capacity building and technology transfer. There are currently twelve countries involved (Tanzania, Nigeria, Ghana, Senegal, Egypt, Tunisia, India, Indonesia, China, Chile, Ecuador, Poland), and others in preparation.

At the regional level, SCP activities are based on city and country-level SCP experiences and aim to facilitate the systematic collection, assessment, sharing and dissemination of information; share operational lessons of experience in urban environmental management; promote the pooling of scientific and technical resources; and support the creation of joint agendas and coordinate programs of action. Professionals from one city project help in start-up SCP projects in other cities; decision makers and stakeholders from SCP cities are increasingly invited to share their experiences at regional (and global) meetings. There are city networks in Africa, Asia, Latin America and the Arab States, aiming to share information and lessons learned as well as to pool expertise and other technical resources.

At the global level, the SCP activities are also driven by the needs and lessons of local demonstrations. The task of global level support is to catalyze capacities of the UN system for local, national and regional capacity building in urban environmental management. Global level SCP activities have steadily expanded from program design and initial resource mobilization, in 1991, to the considerable volume of work associated with coordinating, operationally supporting, and administering the US$ 20 million program in 1996.

The activities fall into six categories, namely: operational support at city, national, regional levels; development of urban environmental management tools; networking among cities and international programs; information and awareness building, and program resource mobilization and management. There is a core team with partners at city, national, regional, and global levels, aiming to backstop activities at city, country, and regional levels; compile lessons of experience and best practice; and develop reusable tools and procedures. Typical issues include water resource management, air pollution and urban transport, environmental health risks, flooding and drainage, unstable slopes, industrial risk, solid waste management, urban agriculture, recreation and tourism resources etc.

In institutional terms, the SCP is a vehicle for inter-agency cooperation. SCP demonstration projects provide agencies and organizations wishing to work in the field of urban environmental management with a framework for their efforts and significant economies of scale. The flexibility of the SCP approach makes it possible to accommodate a wide variety of such support, and the operational structure of the SCP ensures proper coordination of efforts at the local level. The SCP is now working with close to twenty multi-and bilateral partners worldwide, as well as with national and international NGOs and associations of local governments. Currently, most inter-agency collaboration takes place at the city level. Basic funding for demonstrations activities comes primarily from the participating cities, from UNDP country programs, and from bilateral partners such as Denmark, Canada, France, and Italy. Several examples of the inter-agency cooperation should be mentioned:

1 Denmark, the Netherlands, and France have supported rapid assessments and initial project preparation in Accra, Ibadan, Concepcion, Madras, Jakarta, Guayaquil, Katowice, and Tunis.
2 City environmental profiles have been prepared with the Habitat/UNDP/World Bank Urban Management Program in Accra, Jakarta, Katowice, Tunis, Madras, Dar-es-Salaam, and Ismailia.
3 A demonstration module for city environmental consultations has been prepared in Dar-es-Salaam with technical support from Germany.
4 Canada cooperates with the SCP in Dar-es-Salaam on urban agriculture (IDRC), Madras, and Katowice (ICSC).
5 Approaches for addressing urban environmental health issues are being developed and applied with WHO in Ibadan.
6 ILO collaborates on community based infrastructure development and maintenance.
7 Sweden has supported the use of GIS and satellite remote sensing in pilot cities Dar-es-Salaam and Ismailia.

8 Coastal area management issues are being tackled in collaboration with UNEP/OCAPAC in Dakar and Conception, water and sanitation issues with UNEP/IETC in Shenyang, Katowice and Conception.

9 Industrial pollution and risk concerns are being addressed with UNEP/IEPAC in Dakar, Conception, and Katowice.

10 The Netherlands are supporting high priority activities at the global level such as urban environmental management tools and networking among SCP partner cities.

City-level successes achieved to date have shown that the SCP is an excellent facility for inter-agency collaboration, through which diverse technical and financial resources from external support agencies can be mobilized and integrated. In particular, because of the uniquely complementary resources of UNEP and Habitat, the SCP is also an ideal vehicle for institutionalizing collaboration of the two agencies in the urban environment field. Among the Habitat/UNEP Supplementary Core Activities for 1996-1997, supported by supplementary core funding from contributing partners beyond Habitat and UNEP (for example, the Netherlands, Denmark) the following projects should be mentioned: tools for remote sensing and GIS technology; tools for air quality management; tools for environmental risk assessment; tools for addressing gender issues in sustainable urban development; and tools for conducting city consultations.

Other core projects include CD-ROOM and UNEPnet home page; non-recurrent technical publications (approximately five per year); videos documentaries; and EPM indicators. Recurrent activities included annual meetings of the SCP network cities; "Citylink" e-mail communication among SCP cities via UNEPnet; operation and maintenance of home page on UNEPnet; "Sustainable City News", a quarterly newsletter; and routine response to queries.

Conclusion

As I have argued, the SCP is a joint UNCHS/UNEP facility. In the past years several UNEP programs (including the International Environmental Technology Center-IETC, the Global Environment Monitoring System - GEMS, the Industry and Environment Office-IEPAC, and the Oceans and Coastal Areas Program-OCAPAC) have implemented joint projects with Habitat in SCP cities. Several other joint activities are under preparation. Based on these developments, the directives for joint programming given by the Governing Bodies of Habitat and UNEP, and the growing demand for the type of support the SCP can provide, the executive heads of both organizations have decided to strengthen and expand their cooperation in the SCP.

By the end of 1995, program-level cooperation was formalized through an agreement between the two United Nations agencies that includes shared SCP core funding (to cover the costs of an expanded core team, facilities, duty travel, communication and other basic requirements) and joint trust fund arrangements for supplementary funding (to channel funding from other contributors for the development of management tools, and other global level support). Supplementary funding already received will support a wide range of high priority core projects.

Given the strong donor support, expanding city demands, and reconfirmed partnership between the two United Nations agencies responsible for urban and environmental management, the new Sustainable City Program continues to hold extraordinary promise for advancing sustainable urban development and participatory local governance.

Finally, it should be stressed that the Sustainable Cities Program supports a learning-based process to advance collective know-how among the partner cities. Disillusioned with the traditional master plan approach to urban management, cities throughout the world are searching for new approaches to governance and administration. Prominent among the lessons learned are the following:

1 Mobilization of local resources is more effective than reliance on external support.
2 Broad-based stakeholder involvement is more effective than master-planning.
3 Bottom-up problem solving is more effective than top-down decision-making.
4 Need-driven strategies and actions are more effective than supply-driven studies and plans.

10 Innovative strategies for sustainable development in the Middle East and North Africa

Christian Arandel*

Introduction

The Middle East and North Africa have been at the cross-roads of world trade, finance, and culture for centuries. It is an area that has traditionally displayed a strong entrepreneurial spirit, with business tightly interwoven into its social fabric. However, over the past forty years, several factors have stifled the region's historic business drive. A move towards highly centralized, government-controlled planning has edged out private initiative and hindered economic development. Large and inefficient public enterprises have become fiscal burdens on governments in the region, curtailing their financial ability to develop human resources through quality health care and education. The discovery of oil and the instant wealth it represented became a golden crutch that boosted the limping economies of the area. The region's prosperity no longer rested on a solid base of entrepreneurship, a limitless and even expandable asset, but instead, flowed mainly from the sale of an exhaustible natural resource.

In the current global economy, where change is rapid and where human resources are becoming increasingly central to the prosperity of national economies, the countries in the Middle East and North Africa need to adopt innovative strategies, different than those they have relied on in the recent past, if they are to reclaim their place on the world economic stage as credible, competitive players. Drawing on the historic spirit of enterprise that has shaped the region for millennia is their best hope for achieving that end.

Because urban centers represent such fertile ground for bold and creative ventures, cities can lead the way in pursuing sustainable development. The region's cities are efficiency devices that bring goods, services, and markets together in a vibrant and interactive social environment. They are where two of the area's most important assets are concentrated. The first of these two strengths is the entrepreneurial potential and ingenuity of the poor, whose businesses escaped the suffocating confines of government planning and are now where the opportunities for economic growth reside. The second asset is the region's rich cultural and architectural heritage, embodied in the historic centers of its cities, once economic hubs that linked the region's economy to rest of the world, but now largely neglected. Both of these assets represent resources that have all too often been overlooked by policy makers and development planners. However, if carefully nurtured, these resources can have a significant impact on the region's social and economic health.

The following is a concept paper developed by Environmental Quality International (EQI) in cooperation with the Urban Management Program for the Arab States Region (UMP-ASR). The UMP-ASR is a joint initiative of the World Bank, the United Nations Development Program, and Habitat, and EQI is the regional support office for the program in the Middle East and North Africa. We have identified three areas where focused activity would help mine the promise of the region's entrepreneurial spirit and its rich history manifested so eloquently in its cities. They are: the expansion of small businesses, the promotion of informal sector participation in the environmental services industry, and the revitalization of historic city centers.

Small business support provides a sound platform for generating employment opportunities and creating a strong, diversified urban economy; the promotion of informal sector participation in the environmental services industry capitalizes on the entrepreneurial capabilities and resourcefulness of the poor to reduce the cost and increase the efficiency of municipal services; and the revitalization of historic city centers offers a major opportunity for increasing the region's market share of tourism, one of the fastest growing industries in the world. All three areas draw on the vitality of the private sector in stimulating the urban economy and improving urban management practices. In this paper, we offer a thumbnail sketch of the context for activities in those three domains, and then outline our strategy for development in each area.

Additionally, we also introduce gender and the media as two factors to which careful attention must be paid if the approaches summarized for three focus areas are to have significant and sustainable impact. Taking gender into account at all stages of a project's cycle helps ensure that women can benefit from, and contribute to, development efforts in equal measure to men. Engaging the media with the issues of urban and environmental development injects them into the public domain, causing changes in policy to address contemporary needs and future challenges.

Finally, through this paper, we also attempt to show that a shift in thinking and a change in the way the region has perceived itself over the past few decades is necessary if it is to break out of the bind of under-development. We endeavor to demonstrate the importance of redressing social inequity, to underscore the value of rekindling the spirit of entrepreneurship which has thrived for centuries in our streets and public markets, and to demonstrate the utility of involving the mass media as mediators in the debate on needed social and economic reform.

New franchises for sustainable development

Background

Supporting the entrepreneurship of poor people is central to promoting sustainable development and promoting broad-based urban economic growth in the region. Many of the poor are small and micro enterprise owners and their businesses represent the mainstream of private sector economic activity in the region. In Egypt, for example, they account for somewhere between 75 per cent and 90 per cent of private non-agricultural employment. While there is definitely substantial variety among small and micro-scale businesses in the region, there are certain characteristics that they share. They tend to be labor intensive operations that use basic technology. They also have quick start-up times, are remarkably innovative, and have shown themselves to be extremely agile in responding to shifts in market.

While their access to new technology is generally limited, they are innovative enough to incorporate appropriate technology whenever it is made available to them. Furthermore, these operations provide products and services to primarily low-income communities at affordable prices, a market often misunderstood and rarely reached by larger companies. And finally, more than any other segment of the economies in the region, small businesses show the outstanding potential for job creation, particularly among the poor, a very important attribute given the pressure of the region's rapid population growth. When small businesses receive the right support, they flourish, and as they do so, they spark other enterprises, infusing urban economies with much needed momentum.

Despite the enormous potential of this sector, its growth is significantly handicapped by the paucity of financial services, notably savings and loans, available to small business owners. These entrepreneurs consistently list their limited access to credit as a major obstacle to the growth of their businesses. In a 1992 survey of small business owners in Egypt, over 95 per cent of the respondents singled out a lack of working capital as either the most important or the second most important constraint faced at business start-up. Small businesses face this scarcity of credit because banks have traditionally been structured to meet the needs of medium and large business and high net worth individuals. Furthermore, organizations that could serve small business owners are unequipped for, or constrained by, regulatory frameworks from mobilizing the savings of the poor, an important source of funds that could be recycled back into low-income communities as loans. No matter how much potential small businesses have, these engines for economic growth can go nowhere unless they have access to fuel. Consequently, one of the most central elements to promoting urban development in the region is to make financial services available to individuals out of whose reach they have traditionally been.

Providing credit to small businesses fosters their growth, but it is also an area that can be very lucrative for those who offer the financial services. Investors in this field can set interest rates at level that will yield very attractive returns. As illustrated in the following example documenting the daily transactions of a street vendor in Alexandria, Egypt, micro finance providers have a wide range within which they can set their interest rates while still remaining well below those currently endured by the small scale borrower. The vendor obtains his fish supply free of charge from his wholesaler, and settles his accounts with him at the end of each day. For this privilege, the wholesaler charges him LE 12/kilo for fish that would normally sell at LE 11/kilo. This premium of LE 1/kilo is equivalent to an annual interest rate of more than 3,000 per cent charged by the wholesaler. Since the prevailing interest rate charged by the Bank of Alexandria is 20 per cent, the latitude that small business support programs have in pricing their services is enormous.

Furthermore, repayment installments for small loans are relatively insensitive to interest rate increases. For example, doubling the annual percentage rate of 20 per cent charged by the Bank of Alexandria on a loan of US$ 1,000 to be paid back over a period of 60 days, 10 weeks, or 12 months, will only result in a slight increase in the amount of the installments: US$ 0.29 on daily installments for 60 days, US$ 2.15 on weekly installments for 10 weeks, and US$ 9.84 on monthly installments for 12 months. Finally, the market for small loans is vast and virtually untapped. Institutions that currently supply credit to small business only reached a minuscule fraction of the market. For example, in Egypt, the country with the best record in the region for small business support, micro finance distributors presently in operation have only met somewhere between two and five per cent of the market demand.

The micro finance organizations that realize the largest returns on the loans they extend, and that consequently display the brightest prospects for long-term sustainability, are the ones that have adopted a private sector management style. Through our extensive experience with foundations that extend loans to small businesses, we have found a strong presence of the local business community on the Board of Directors and the integration of several private sector techniques in the running of the organization results in an operation that is agile, efficient, and productive. The Alexandria Business Association, a micro finance provider that, with the assistance of EQI, began extending small loans, is an excellent illustration of this principle: having incorporated many private sector techniques into its methodology, the association now represents one of the most impressive small and micro enterprise support projects in the region.

Approach

To reach a larger share of the small business sector, we propose supporting the creation of franchises that extend loans, and possibly provide other financial services and technical assistance, to small businesses. Launching micro finance initiatives offers investors the unique opportunity to make a significant profit while at the same time fostering the growth and expansion of small businesses. In addition to establishing new organizations for the purpose, an opportunity for the creation of such franchises lies in incorporating micro finance operations into already existing institutions. It represents an attractive option for many interested in expanding their markets and developing new sources of revenue.

Candidates for that type of expansion are numerous. Among them are community bank branches, whose traditional responsibilities are being curtailed as banks automate and centralize their operations, and local branches of the postal service, often already repositories for the savings of the poor. In Egypt, this latter option is particularly appropriate because the savings deposited at those branches endow them with a ready pool of funds that can be injected back into the urban economy as loans to members of low income groups. Institutions that can mobilize savings show tremendous potential for sustainability because the savings collected provides them with a very economical source of funds to extend as loans, allowing them to benefit from a wide profit margin on their services.

A significant deterrent to investors interested in setting up a franchise to extend small loans is the amount of time required to get a micro finance venture fully operational. A review of the expansion of several small business support programs reveals that they typically require one to two years to lay the groundwork for project implementation. Currently, there are organizations, EQI among them, who do assist newly created and already established institutions in setting up the mechanisms and imparting the skills required to extend loans to small businesses. However, for each new client, their services are painstakingly redesigned and remolded, and therefore, do not significantly shorten the initial start-up phase. In a sense, they provide the equivalent of "haute-couture" in systems for loan extension, molded to each client's shape and idiosyncrasies.

To address this liability of time and encourage both the proliferation of new micro finance providers and the widespread incorporation of micro finance services into existing institutions, we are interested in formulating a standardized - "ready-to-wear" - consultation package that can be widely and easily distributed to interested parties. Such a package would include the necessary management skills, loan extension methodology and information systems to allow any organization to begin disbursing loans effectively and profitably. It would allow institutions, new and old, to become completely functional micro finance providers in a matter of months instead of years, and to accrue a profit soon thereafter. Through our in-depth experience in creating new organizations for credit extension and restructuring existing ones to accommodate small business support as an additional activity, we have identified the specific components that make for a solid credit extension program and that should be incorporated into the package. The credit delivery toolkit that we propose will result in substantial efficiency gains in the transfer of skills to candidates for micro finance provision and in the process of loan disbursal.

The marketing, distribution, and installation of the credit delivery toolkit will be based on the model developed by hotel management companies. Those companies offer investors interested in establishing hotels, but lacking the requisite capabilities to run them, the management team and skills to turn the hotels into money-making operations. While they charge investors an initial fee, most of their compensation is performance-based. The experience of these management companies reveals that the ownership of the establishment has less bearing on its profitability than does the management approach with which it is run. This dynamic has intriguing ramifications for the conversion of bank and postal branches to micro-finance providers: even though it may appear that public services may not be the most appropriate micro finance providers, public sector ownership will not significantly dampen the high rates of return on financial intermediation for small and micro enterprises if a private sector management style is embraced. Consequently, a key element of the proposed packages will be to develop a management culture based on the private sector practices that made the success of Alexandria Business Association possible.

The Alexandria Business Association and the Egyptian Small and Micro Enterprise Development Foundation

A review of different venues for the disbursement of credit to small and micro enterprises underscores the importance of private sector qualities to the success of micro-finance operations. The experiences of two micro-finance ventures in Egypt show that program effectiveness and sustainability hinges on the adoption of a private sector approach. Both organizations are non-profit NGOs, one located in Alexandria and the other in Cairo. The Alexandria Business Association (ABA) was originally a business association made up of entrepreneurs from the local community and has had a fairly pronounced private sector dimension since its inception. The Egyptian Small and Micro Enterprise Development Foundation of Cairo (ESED), on the other hand, has only recently made the necessary changes to its management structure to incorporate private sector strengths.

The foundations each received a collateral fund from the United States Agency for International Development (USAid) against which they borrow money at commercial rates to on-lend to their clients: in 1990, ABA was endowed with US$ 8,000,000, and ESED with US$ 2,650,000. They also received separate grants to pay for the expenses of management and operation until they could cover those costs from the revenue generated through loan recovery: ABA received US$ 2,000,000 and ESED was awarded US$ 1,407,300. Both institutions espouse a similar methodology for loan disbursal which circumvents the traditional barriers to financial services: they extend short-term loans for use as working capital, require minimum collateral, offer flexible loan repayment terms, and provide technical assistance to their clients.

Additionally, the organizations send employees directly into the communities they serve: these extension officers identify clients, assess their needs, and monitor and collect loan repayments. This methodology, characterized as a typical NGO approach to micro finance because of the high level of client-extension officer interaction, has proved remarkably effective when combined with a private sector management style.

ABA has been run by a Board of Directors consisting of members from the local business community since the project was initiated. This has resulted in an operation that is particularly productive and sustainable. Several indicators point indisputably to its effectiveness. The foundation has served approximately 21,000 clients, achieving a range of outreach comparable to some of the most successful micro-finance ventures in the world. Its loan portfolio is diverse, including loans to SMEs involved in manufacturing and processes, trade and commerce, and the service industry. The rate of increase in the number of borrowers, data that demonstrates the level of client demand for services and their degree of satisfaction, is quite impressive at the foundation, hovering at around 138 per cent a year. They have achieved a high loan repayment rate at 99.2 per cent, a critical factor in the financial sustainability of the organization.

The level of productivity of the foundation's loan officers, an accurate gauge of the organization's effectiveness given their significant responsibilities, is strong, averaging at 100 clients per credit officer. The means used by the foundation to attain such a laudable level of extension officer productivity illustrates its private sector flavor: loan officers are given a basic salary supplemented by incentives depending on the number of clients maintained and the loan repayment rate secured. Because of the foundation's effectiveness in all of these areas, it began covering its operating costs form the revenue generated through loan recovery in 1992, two years ahead of schedule.

The track record of the ESED has been mixed, but its experience is a compelling illustration of the relevance of a private sector approach to the success of micro-finance operations. In the early stages of the project, the foundation was directed by a management team with a limited business culture, and it was saddled with the consequences of decisions that were not made with profitability of the operation as the first priority. Some of the less than ideal management practices included micro management by the NGO's Board of Directors and the Executive Director's ensuing lack of authority, hiring decisions that were not made with the productivity of the organization as the primary consideration, and an inappropriate distribution of responsibility among staff so that some employees were under-utilized while others filled more than one role. A reluctance in the upper echelons of management to streamline organization structures and update the NGO's information systems also undermined the foundation's efficiency. This management approach translated into performance indicators that were somewhat sobering. Two years into the project, loan officers were reaching an average of 57 clients each, and only 748 SMEs had been served. The level of delinquency on loan repayment was high, at 17 per cent.

In early 1994, the foundation experienced a change in management personnel and style. More private business representation was integrated into the Board, a former banker was hired as the new executive director, and a strategic plan to turn the foundation into a profitable operation was developed. The plan focused specifically on improving extension officer productivity and reducing loan delinquency and led to several fortuitous changes for the micro-finance provider. The organizational structure was modified so that more responsibility was delegated to the Executive Director and his subordinates, and the bureaucracy of the foundation was trimmed through the abridgment of paperwork and the increased use of automated record-keeping.

An incentive system to improve the productivity of extension officers, similar to the one in use at ABA, was instituted. These private sector style changes to the management structure have resulted in a dramatic increase in performance. The foundation loan officers now serve an average of 116 clients each and they secure a repayment rate of 98 per cent. The total number of small and micro entrepreneurs who have borrowed from the NGO has grown to 23,450 and the rate of increase of those borrowers has shot up to 200 per cent.

Informal sector participation in environmental services

Background

For many cities in the region and throughout the developing world, dealing with the environmental costs of rapid growth and urbanization represents a phenomenal challenge. This is particularly true in the area of solid waste management: as cities generate an ever-increasing volume of garbage, the effectiveness of their solid waste collection and disposal systems are declining. In urban centers throughout the developing world, less than half of the solid waste produced is collected, and most - that is, 95 per cent - of that amount is neither contained nor controlled. It is either indiscriminately thrown away at various dump sites at the periphery of urban centers, or at a number of so-called temporary sites, typically empty lots scattered throughout the city. Those open landfills often have environmental impacts that extend beyond their boundaries, polluting nearby water sources and serving as breeding-grounds for disease-bearing rodents and insects. Additionally, the garbage that spills out of these sites impairs the operation of infrastructure systems, destroying pavements and street surfaces and hindering pedestrian and vehicular flow. Furthermore, as existing sites are filled to capacity and new sites are increasingly hard to find, the costs of disposal are sharply on the rise exacting a huge toll on already strained municipal budgets.

Many urban centers in developing countries have two parallel, but separate, systems that handle the cities' solid waste. The first is the formal one administered by the government: it tends to be costly and inefficient. The second is highly informal, and essentially involves communities of scavengers that compensate for the slack in municipal services by collecting, sorting, recycling and selling waste. The scavengers, often the cities' most physically, socially and economically marginalized inhabitants, recognize the potential value of certain materials, such as plastic, paper, tin, and bones, and turn to the recovery and marketing of that refuse as a source of income. Cooperation between those two systems is exceedingly rare. However, in Cairo, where the government entered into partnership with the informal garbage collectors, the fusion of the two systems was successful on many levels. The experiment offered new insight into how this merger could increase the efficiency and reduce the cost of street cleansing and waste disposal while creating work for the urban poor.

Because the government formally integrated the garbage collectors, called Zabbaleen, into the city's waste management system, they were able to improve and expand their solid waste collection. The Zabbaleen were organized into small franchises, a move that allowed them to provide adequate service without a dramatic modernization of technology. Furthermore, with the help of an urban upgrade program and the provision of credit in the Zabbaleen settlement, many of the garbage collectors established small businesses for the processing of recycled trash, allowing them to increase the profit that they gleaned from the recyclable waste. Interestingly enough, once a business model was introduced into the community, several other enterprises sprung up, businesses having to do with industries other than the processing of recyclable garbage such as the production of merchandise from recovered waste, trade and commerce, and service.

Approach

Given the success of the Zabbaleen experiment in ameliorating waste management services and in developing the entrepreneurial potential of the poor, we would like to replicate the Zabbaleen model and experience in other cities throughout the region. The approach we envision is two-tiered: the first phase involves formally integrating the informal sector in municipal solid waste management, and the second consists of small business promotion in scavenger communities.

To integrate the formal and informal systems of solid waste management, we see the need for several steps. The first step is basically to introduce the concept to both the government, who may be skeptical that any relationship with informal garbage collectors will be truly synergistic, and to the scavengers, who, as they generally exist on the margins of legality, are likely to be suspicious of any government authority. As part of this process, specific measures to reduce the poor's cost of entry into the solid waste collection trade must be presented to all parties involved, including the municipality, the scavengers and relevant community organizations. These measures include, among others, parceling up the city into small service units, as small as a few city blocks, that the scavengers can feasibly bid for, or the abolishment of bidding packages altogether. The initial presentation of the idea must supported by the second step, which is the provision of the required legal and technical support to actually put a working partnership between the government and scavengers in place. As a third step, the garbage collectors should be provided with the necessary training and technical assistance to ensure that they can operate their solid waste collection and recycling franchises as profitable operations.

The second phase concerns the establishment and support of small recycling businesses in the scavenger settlement. This phase ensures the sustainability of the association between the formal and informal solid waste management systems by supporting it with a broad and growing economic base for the marginalized community. As with the first phase, the second stage in composed of several important steps. The first involves a thorough investigation of the income-generating opportunities in the recycling industry accessible to the garbage collectors along with a careful market analysis to determine the location and size of the demand for recycled waste. The second step is a more direct intervention to promote small businesses, that includes introducing appropriate technology for the processing of garbage to interested entrepreneurs in the community and supplying them with credit to purchase the machinery or to make other investments. The efforts of this second phase must be complemented with a comprehensive upgrading program for the community because the abysmal infrastructure of scavenger settlements not only impacts the health and living standards of the inhabitants but also hinders the growth of small businesses. Few machines can function without electricity and water is essential to many waste processing activities.

Harnessing the power of the informal sector: the case of the Zabbaleen

In Cairo, responsibility for the management of the solid waste system is currently shared between the Cairo Cleaning and Beautification Authority and a traditional private-sector waste collection system that has evolved over the last fifty years. In the traditional configuration of this system, the Wahis, or "people of the oasis", served as brokers and administrators, while the Zabbaleen actually collected the city's garbage. The latter were originally landless agricultural laborers who migrated to Cairo from Upper Egypt in several waves beginning in the 1930s and 1940s. Because they had little education and were equipped with few technical skills, they had turned to garbage collection for their livelihood, purchasing the right to collect waste from the Wahis, who had been involved in the trade since the turn of the century. The Zabbaleen were able to use the organic waste as pigfeed and earn limited profits by selling pork products. They complemented this revenue with the sale of sorted recyclable solid waste like bones, glass, plastic, and paper.

This system served Cairo relatively well until the mid-1970s, when it began to fray under the pressure of the capital's explosive growth. The fragmented character of the workforce and the informal nature of the work was limiting the system's ability to meet the needs of the rapidly expanding city. Although, at one point, the municipality seriously considered replacing the traditional system with a modern, mechanized one, it was eventually responsive to policy initiatives to integrate the Zabbaleen into Cairo's waste management system. In cooperation with EQI and two NGOs working in the scavenger community, the Association for the Protection of the Environment and the Garbage Collectors Association, the municipality opted for a franchise arrangement in which the Zabbaleen and Wahis were organized into more than 80 small independent companies, each responsible for a terrain of about 500 households.

This solution not only enabled the Zabbaleen to continue collecting Cairo's garbage and survive economically, but it also introduced to this group a business framework that would serve as the basis for the development of various cottage industries related to solid waste. The limited service areas meant that they did not have to invest a large amount of capital to upgrade their technology, but could build on their skills and experience instead. Additionally, because the Zabbaleen were generally assigned the neighborhoods that they had historically served, they were able to increase their prices to cover the small additional organizational cost of becoming franchises without jeopardizing their client base. The Zabbaleen franchises soon began to display the characteristics typical of vital small businesses: they were labor intensive operations that used basic technology, cost effective and efficient in delivering their services, and responsive to their markets.

The division of Cairo into small bidding packages was supplemented by several projects designated to improve the living conditions of the Zabbaleen. The Governorate of Cairo targeted the Manshiet Nasser Settlement, which was, and still is, home to about half of Cairo's garbage collectors, as a significant beneficiary in its urban upgrading program of the late 1970s. The settlement was included for two reasons: first, because it was thought that the improvement of living conditions in the settlement would allow the Zabbaleen to provide a better quality of service; and second, because the upgrading needs of the settlement were particularly urgent. At that time, the settlement lacked water, sewerage, electricity, and roads. The garbage collectors had few resources to build permanent housing structures and most lived in homes made of low-grade steel sheeting. The settlement was choked with garbage and the health standards were abysmal.

The upgrade program, though badly needed in the settlement, lacked the capacity to ensure that an improved standard of living for the Zabbaleen would be maintained over the long term. What was missing was a strategy to foster sustainable economic growth in the community that would increase the family incomes of the Zabbaleen and support the new franchise arrangement for garbage collection. After a study tracing the lifecycle of the garbage collected demonstrated that the Zabbaleen would sell the sorted waste at very low prices to recyclers who, after some limited processing, would resell it at a substantial profit, the idea of supporting industries among the Zabbaleen that would augment the value of the waste collected began to take shape.

The promotion of those industries among the Zabbaleen took the form of the Small Industries Project launched in 1983. EQI worked with a handful of garbage collectors to help them set up small-scale waste recycling businesses. The organization exposed the entrepreneurs to low-cost technological innovation for waste recycling, such as the technology needed for plastic granulation and composting, and loaned them the capital necessary to purchase the new equipment and get their businesses off the ground. It followed through with concerted technical assistance for the proper use and maintenance of the new machinery, and linked the entrepreneurs to markets for their goods. When other members of the community observed the success of these half-dozen initial recycling modules, several other Zabbaleen sought loans and technical assistance through the project to establish similar businesses.

However, one of the most remarkable effects of the project was the mushrooming of like industries within the settlement that neither sought nor received the project's financial backing. While prior to the project, not a single garbage collector owned a machine for processing recyclable waste, observation of the successful performance of the machines and the profitability of the modules led others in the community to risk their own money or money borrowed through community networks to set up their own workshops, often as direct copies of those already in operation.

The impact of the project has been dramatic: before its inception, not only was no recycling carried out in the settlement but only a bare minimum of services were available. However, by 1993, only ten years after the project began, 215 flourishing enterprises had emerged. Of these, 42 per cent were commercial establishments, 36 per cent industrial, and 22 per cent service-related. Furthermore, a number of the manufacturing enterprises went beyond simply recycling and began using the waste as raw material to manufacture plastic goods such as clothes hangers and toys. Because these industries were able sell their products at very reasonable prices, they met the needs of a large market of low-income consumers who could not afford merchandise not made from recovered waste.

This rapid spread of industries in the Manshiet Nasser Settlement suggests an important method for generating sustainable and self-perpetuating development. If a development initiative makes good business sense and is widely accessible, it has a way of spreading like wildfire. All the Small Industries Project did was introduce a business model into the community, and prove that the industries they were supporting were profitable. That was all it took to motivate others to shoulder the perceived risk in setting up a business. The case of the Zabbaleen indicates that development organizations have the most profound and lasting impact when they provide the initial push for economic development, and then allow their efforts to grow, multiply, and expand driven by their own momentum. In other words, development efforts are the most effective when they seek to facilitate rather than engineer economic growth.

The proliferation of informal business related to solid waste collection, recycling, and reuse combined with the government upgrade efforts has resulted in a striking transformation in the standard of living in the settlement. Whereas in 1981, there was no infrastructure whatsoever, by 1993 most of the roads were leveled, and the large majority of houses had access to potable water and electricity. These changes have been accompanied by an improvement in the health status of residents as a group thanks to better sanitation, improved hygiene, and the various community-based health programs initiated at the settlement. The infant mortality rate in the settlement is a strong indicator of this improvement: the rate dropped from 240 per thousand in 1979 to 117 per thousand in 1991.

While EQI has completed most of its activities in the settlement, the Association for the Protection for the Environment and the Garbage Collectors' Association have adopted a more prominent role in promoting community development activities. They have expanded their community health and literacy programs, and they continue to actively nurture the proliferation of cottage industries related to recycling. Their commitment and the community's involvement in the projects they support is another indicator that the changes in the settlement are ones that will be sustainable over the long term.

Regeneration of historic city centers

Background

The Middle East and North Africa region has been called the cradle of human civilization. Early in its history, the region developed vibrant cities that are now among the most ancient and beautiful in the world. The historical remains in the cities stand as testimony to the various civilizations that emerged and flourished in the area. They also speak to the vitality of those urban centers where order was maintained through advanced legal systems, and where scientific knowledge and culture drew people from all over the world, laying the foundations of elemental sciences as mathematics, astronomy, and medicine. Examples of thinkers who heralded form these cities are Euclid, the famous mathematician who was born, lived and died in Egypt; Ibn Khaldoun, the founder of social sciences; and Ibn Sina, a pioneer in the field of medicine.

The urban tissue and architecture of Arab cities bear witness to the region's long and complex history during which different peoples often crossed each other's paths, their cultures mingling, sometimes with struggle, but mostly with grace. Furthermore, they are a testimony to the extent to which the region contributed to, and benefited from, the

177

civilization of the rest of the world: they are an eloquent reminder of the world's shared heritage. Undoubtedly, historic cities that dot the region today represent one of the area's greatest assets.

Unfortunately, this rich heritage has been and continues to be threatened by a set of inter-related forces which have marginalized historic city centers and led to their inexorable decay. Since the turn of this century, cities in the region have endured successive and important waves of rural-urban migration compounded by the pressure of rapid population growth. For historic centers, this phenomenon most often meant overcrowding, increased strain on infrastructure unable to accommodate the new levels of use, and eventually, economic decline. The original inhabitants of historic areas moved to modern neighborhoods and were replaced by impoverished rural migrants. The new settlers had neither the resources nor the awareness to preserve the architectural heritage in their living environment. In addition, as various new industries relocated into the city centers, they supplanted the city's traditional industries and crafts as a source of livelihood for its inhabitants and often had a negative impact on the local environment. Many residential buildings were transformed into small workshops or warehouses, changing the character of some areas.

Historic districts were also marred by ill-conceived planning approaches. In an attempt to meet the space and infrastructure demands of the expanding urban population and economy, Arab cities engaged in a process of modernization that included rapid construction with little consideration for the heritage of urban centers. The use of central planning and other engineering techniques, applied with little regard for social and cultural considerations, meant that roads were cut through historic districts, destroying many historic buildings and rending the physical and social fabric of the old city centers. In addition, traditional architecture was replaced by a new, low-quality architecture imported from abroad. Not only did this architecture often clash with existing buildings, but it was poorly adapted to local environmental conditions and ways of life.

Even as municipalities were making significant changes to the city centers in an effort to manage the swelling urban population, an important source of income for the city centers was overlooked. Tourism, one of the fastest growing industries in the world, is significantly under-developed in the region. While by some estimates, travel and tourism represents more than 10 per cent economic output, investment, and employment worldwide, it accounts for only eight per cent in the Middle East and North Africa. The rich history and culture recorded in the region's city centers and transmitted through its artisans could be more assertively developed, an resource that can fortify the economic base in the heart of Arab cities.

Approach

Transforming historic cities from relics of the past into powerhouses of economic and social development is one of the most daunting challenges facing Arab cities today. The regeneration of historic cities requires the adoption of new and bold measures. Traditional approaches that focus too exclusively on historic preservation without addressing the root causes of decay represent a wasteful use of resources, as monuments will decay again and again, requiring never-ending restoration. Preserving these areas implies more than restoring buildings and preserving their architectural heritage; it means that the physical upgrading of historic centers must be complemented with strategies to improve the socio-economic conditions of the people who live in those areas. In other words, these cities must be revived and rehabilitated to interact more productively with newer urban areas: city centers should be turned into active cells within the urban body, rather than only beautiful ornamentation. In order to achieve this comprehensive objective, we see the need for action at the regional, national and local levels.

At the regional level, we would like to facilitate the transfer of innovation. Several cities in the region have had successful experiences in the regeneration of historic cities and an important pool of knowledge and expertise is now available. In particular, a few countries have been able to make great strides in historic revitalization and have demonstrated a capacity for innovation. Tunisia is an example of one of these important successes, having recognized early on the value of its urban heritage as a source of pride and economic development. Disseminating Tunisia's experience in promoting the use of historic buildings for contemporary purposes and in creating a policy and regulatory framework for the regeneration of its medinas would provide other Arab countries with an excellent model for their own efforts.

At the national level, we intend to foster a national discussion between decision makers, experts and community members on the revitalization of historic cities. The main venue we plan to use are consultations during which the people involved can share their concerns and experiences, and possibly outline concrete measures to regenerate city centers. However, we also plan to encourage the media to cover the issues concerned so as to include the general public in a wider debate. An important issue to be addressed in this national discussion is how to promote legal and policy environment that is conducive to protection and regeneration of historic districts. For example, legislation designed to protect historic monuments in Egypt does not allow for the use of historic sites for contemporary purposes; according to the terms of current laws, historic city centers are to be preserved so that they can serve a purpose similar to that of a museum. A second issue that should be considered in any national discussion is the adoption of building standards and regulations which are specifically tailored to old districts and which guarantee that new constructions or renovations respect and strengthen their identity.

At the local level, we want to support the design and implementation of revitalization projects that rest on three different pillars: adaptive re-use, small business development, and community participation. Adaptive re-use refers to the use of old buildings either for their original purpose or for a new purpose compatible with the original one. It is an effective means to integrate these buildings into the social, cultural and economic structure of the city. Through adaptive re-use, an old caravanserail can take on a new life as an hotel, a cultural center or a center for handicraft development. Furthermore, re-using old buildings offers opportunities for partnership between local residents, local and central governments, and the private sector. Small business development, the second part of the strategy for the local level, is critical to the creation of a sound and diversified economic base in historic districts and will help ensure that economic development efforts in those areas are self-sustaining. Making credit available to small-scale entrepreneurs in historic centers will help revive traditional handicrafts and will encourage the growth of vibrant industries in the heart of the region's cities.

Finally, the participation of the communities living in the historic centers in regeneration efforts is the strongest guarantee that the changes in those districts will be lasting. Involving the inhabitants in regeneration projects strengthens those initiatives for several reasons: firstly, it ensures that the projects are compatible with community needs and priorities, and secondly, it allows them to draw on the residents' experience and knowledge of their surroundings. Thirdly, and perhaps most importantly, it develops the community's sense of ownership over the projects and their responsibility for the maintenance and preservation of their living environment.

The regional, national, and local initiatives described will generate a process of sustainable development in historic city centers that will not only engender an improvement of physical and economic living conditions in those areas, but that will also allow the region's city centers to capture a larger market share of cultural tourism, one of the fastest growing industries in the world. The revitalization of historic cities and the promotion of tourism are mutually re-enforcing. Attracting culturally minded tourists and businessmen and women who appreciate the value of historic buildings and monuments can be magical for the urban economy, expanding business and filling city coffers, providing resources for the upkeep of buildings and the maintenance of infrastructure and public space. Additionally, linking historic centers to the wider urban, regional, and global economies will make them both beneficiaries of and contributors to world economic and cultural growth.

Revitalizing a city's heart: the Medina of Tunis

The *Association de Sauvegarde de la Médina de Tunis* (ASM) was created in 1967 to tackle the formidable challenge of rehabilitating the historic city center of Tunis. A major part of that endeavor involved the difficult task of convincing planners and decision-makers of the value of historic preservation. In the 1960s and 1970s, Tunisia's eyes were turned to the future, not the past; they were riveted on the development of the country and on solving the most pressing social problems identified by planners, namely those of health, education, job creation, and housing.

In this context, the Medina of Tunis, the heart of the old city, was not considered as an asset, but rather as a road block on the highway to modernity. The city center that would later be added to the UN World Heritage lists as one of the most important examples of a traditional Moslem urban center represented for many all the archaic social structures and modes of production the country wanted to do away with at the time, and was a physical obstacle for the planners and engineers who had drafted new maps featuring a renewed and rational urban structure for Tunis. Their plans included the construction of large avenues that cut through the historic city, threatening the intricate maze of small streets and alleyways that were, in a sense, the threads of the Medina's historic urban fabric.

From the outset, the ASM set out to prevent the most irreparable damage. It argued that not only was the Medina worth preserving, but also that it had an integral place in the modern city. In keeping with this perspective, the ASM did not limit its range of actions to the preservation and restoration of historic monuments but rather strove to develop the Medina's role in the city's economy as a vibrant center for business, handicraft, and as an attractive place to live.

In collaboration with the UNESCO, the ASM presented a plan in 1973 for the development of the Medina to the relevant authorities of Tunisia. This plan spelled out policies in three main areas: the protection and renovation of historic monuments, the development of handicraft and organization of trade, and the upgrading of slum areas and the overall improvement of the housing stock. While initially this plan was not given any serious consideration, it was integrated into the Greater Tunis Development Plan in 1980. The components of the initial plan that were finally adopted were mainly limited to zoning regulations for controlling building use and new development. However, the ASM was designated at the agency responsible for all planning in the Medina.

The success of several high profile projects undertaken by the ASM was instrumental to the incorporation of a number of the organization's precepts into the Greater Tunis Development Plan and to their official role as stewards of the Medina. The ASM's urban renovation project in the district of Hafsia is one of these projects and it offers a good example of the association's approach to historic preservation. The Hafsia was originally the Jewish district of the Medina of Tunis, and its poor condition had led to municipal clearance actions as early as the 1930s. The ASM first tried its approach to renovation by rehabilitating a small section of the district in 1973. The success of this early attempt led to planning the renewal of the entire area, after 1981.

ASM's approach rested on two basic principles: that existing housing stock should be upgraded where possible, and that new buildings should be built in an architectural style compatible with the one reflected in existing buildings. With the help of the World Bank, participating for the first time in an urban renewal project, the ASM was able to fulfill its two main goals and develop a cross-subsidization scheme that increased social diversity in the area by attracting financially comfortable residents while allowing the poorest households to stay. Total investment for the project was US$ 14 million, a sum that covered, among other things, the construction of 400 new homes, several commercial premises, a kindergarten, a dispensary, a public bathroom, and a café. The effectiveness of ASM's approach was internationally recognized when the organization won the Agha Khan Architecture Prize in 1983.

In all of their restoration efforts for historic monuments, the ASM has nurtured traditional *savoir faire*, notably stone carving, ceramic tile making, and stucco carving. The emphasis has been put on adaptive re-use, and several monuments have regained a life of their own as cultural centers or offices. ASM have complemented their efforts at housing upgrading, urban renovation and monument restoration with steps to boost the local economy. Specifically, these have involved improving the quality of the craft production and aiding the remaining souks in marketing the products for which each was renowned.

While much remains to be done, the efforts of the ASM and its partners have already reaped substantial rewards. The Medina of Tunis is currently one of the most vibrant historic city centers in the Arab region.

It is more than a frozen testimony to a bygone past; it is a vibrant commercial and handicraft center and a desirable place to live. The mixed crowd of local residents and tourists that stroll the winding streets of the Medina, in search of a good bargain on traditional crafts or a close-up view of historic buildings, is the most visible sign that the city's ancient heart is still beating.

Gender

Background

Because women in the Middle East and North Africa face social, economic, and cultural constraints that are particular to them as women, gender must be given special importance as a critical variable in sustainable development. Indicators such as levels of literacy and employment rates illustrate the discrepancies between the circumstances of men and women in the region. Illiteracy rates for women throughout the area are high, especially in countries such as Algeria, Egypt, Morocco and Saudi Arabia, where more than half the women are unable to read or write. Rates of participation of women in the labor force of Middle Eastern and North African countries are low, rarely exceeding 15 per cent.

Development initiatives have, for the most part, been ineffectual in correcting the striking gender-based inequity in the region. Many development programs skirt the issue of social inequality between men and women altogether, while others subsume the needs and priorities of women under those of the family. Both of these approaches display important limitations when it comes to addressing the concerns of women and contributing to their well-being. The services offered by projects that do not take into consideration the differences in the priorities of men and women are almost inevitably more tailored to the needs and circumstances of men. Many otherwise effective small business support projects fit this profile perfectly: women typically represent only a sliver of their loan portfolio.

An example is one drawn from the case studies presented in this paper. Women clients make up only about 10 per cent of Alexandria Business Association's (ABA) and the Egyptian Small and Micro Enterprise Development Foundation's (ESED) loan portfolio, a proportion that is under-representative of the very large market that women entrepreneurs constitute. The reasons that women are under-served by these credit delivery programs are complex, but many of them have to do with the fact that women-owned businesses tend to be configured differently than businesses owned and managed by men. Many women run their businesses out of their own homes and operate them on a seasonal basis. ABA and ESED, along with many other micro-finance establishments in the region, have not yet developed criteria to evaluate the potential of businesses that do not fit the traditional mold. Furthermore, they lack gender-specific strategies to reach women and to provide them with information about their services. This programmatic weakness may be partially due to the fact that, in both organizations, women make up less than 10 per cent of the loan officers, the employees who have direct contact with the projects' clients.

Projects that equate the social and economic well-being of women with the welfare of the family to which they belong have failed to take into account the gendered distribution of work and access to resources within the family itself. The Small Industries Project at the Manshiet Nasser Zabbaleen settlement of Cairo, Egypt, is a case in point. In the settlement, about 60 per cent of the women are involved in activities that directly generate income for the household, in addition to their other responsibilities of child-rearing, cooking and cleaning, and caring for household livestock. Of those women, about three-fourths are involved in waste sorting, and the remaining quarter are primarily involved in recycling activities. Most of the women receive no financial remuneration for their work: 90 per cent of waste sorters and approximately 75 per cent of women involved in recycling activities secure no direct payment for their efforts. This curtails their economic and social independence.

It may be argued that women benefit indirectly from their activities, as their contribution improves the standard of living of the family as a whole, but this notion is based on the assumption that resources are shared within the family equitably, an assumption that rarely holds true. In 1981, the Small Industries Project began promoting the emergence of small family-run enterprises in the settlement. While the project has generated a steady source of income for a large number of the families in the settlement, it has done little to change the gendered economic patterns prevalent among the Zabbaleen. The development of recycling industries has not only failed to provide women with any direct income, but it has actually increased their unpaid workload.

Over the past two decades, there has been significant a discussion of how to make the process of development more responsive to gender. However, it has borne few results and the fruits of development remain inequitably distributed according to gender. Women's contribution to the national product and family welfare continues to be systematically underestimated, and when acknowledged, the recognition rarely translates into policy. For example, so many studies on women in development have pointed to the positive correlation between women's education and the general health of her family that it has become common wisdom; yet, concerted effective efforts to improve the access to and quality of education for women and girls have been lacking. The importance of women's participation to the success and sustainability of development makes their continued marginalization not just discriminatory, but economically and strategically unsound.

It is important to note, however, that several attempts have been made to address the needs and priorities of women. Unfortunately, these have generally amounted to project modifications through an add-on tactic: methodologies to target women better have been grafted onto existing projects. While this strategy has represented an attempt to demonstrate women's contribution to development and to involve them in the mainstream of the development process, it focuses on women in isolation from the context in which they live. Projects have stretched to accommodate some circumstances that women confront, but fundamentally, the paradigms on which those initiatives are based remain basically the same.

Such attempts reflect a widespread tendency to consider women divorced from the rest of their lives. Gender is often looked at through the narrow frame of specific development projects, and any social constraints, responsibilities, and relationships of women that fall outside limited boundaries of that frame are overlooked or discounted as unrelated to project implementation. For example, a small business support project would most likely disregard the obstacles that women face in trying to secure stable housing as well as the high illiteracy rates among women.

However, lack of access to stable housing often deters women from setting up micro enterprises out of their homes - the place out of which most women-owned small businesses operate - and the inability to read or write can make it impossible for a woman to fill out the necessary paperwork to apply for a loan. The "projectizing" of gender has resulted in a superficial, piecemeal approach to gender where the social concepts and structures that govern how resources, opportunities, and power are distributed in society are rarely tackled. Gender is taken out the public and political domain and confined to the mundane details of project implementation.

Admittedly, it is unrealistic to suggest that every individual development initiative should address all the variables that may discourage women's full participation and contribution. However, it is not only realistic, but imperative to submit that development, if it is to have the most beneficial impact possible, must be wedded to concerted political activism to redress gender inequality.

Political activism for gender equality in the Middle East and North Africa is likely to the most effective if it is based on the complex realities lived by both men and women in the region. As in any part of the world, culture and social attitudes play a significant role in widening the gender gap. Prevailing views on the region argue that religious and cultural currents affect women's lives to a greater extent than do trends of economic change and urban growth. "Cultural restrictions" are often cited as the major, if not the sole, factor accounting for significant indicators of gender inequity such as low participation of women in the labor force or high rates of illiteracy. While culture undoubtedly shapes gender relations, other factors, including political realities and patterns of industrialization, have a significant impact on the distribution of resources, power, and authority between men and women.

A study by economist Valentine Moghadam suggests that low employment rates for women in the Arab world have much more to do with patterns of industrialization than with cultural norms. She argues that female labor participation is low, especially in the formal industrial sector, because the capital intensive industries and technologies tend to favor skilled labor, labor that because of women's limited access to training and education, is typically male. A sophisticated understanding of gender in the region also depends on an examination of the social and economic realities that underpin many of the conservative cultural currents that are surfacing in many countries across the Middle East and North Africa.

For example, sociologist Hoda Hoofar argues that changes in the Egyptian economy has led many middle-class women to reappraise the traditional role of women as homemakers. She points out that, as market ideology is becoming more preponderant and as cash is becoming an increasing valuable type of income, women's non-monetary contributions are less visible and are often dismissed. Meanwhile, their options for finding reasonably paid work are becoming more scarce. To compensate for a balance of power in the family that is shifting in favor of men, many women are embracing traditional doctrines that commend women for staying at home and that therefore provide them with greater social, if not economic, status within the home and the community.

Not only must political activism to redress gender inequality be based on an understanding of women's formal and informal participation in political and economic spheres, but it should also display the courage to confront issues often considered sensitive or inflammatory. Marital violence is one example: surveys conducted in Egypt state that one of three women has been battered by her husband at least once since her marriage and half of those women have been beaten at least once in the past year. Another area that needs attention in family law. Gender discrimination in divorce and custody laws undermines the economic security and autonomy of women.

Political activism, if it is to be effective, has to take on many forms, ranging from small local efforts to significant policy change at the government level. Development organizations, for one, can work to redress gender inequality by moving gender from the category of peripheral concerns, to be dealt with if there are spare funds, to the category of central priorities. This shift requires institutional commitment that is backed up with the allocation of funding and personnel. Furthermore, the development field in the Middle East and North Africa lacks gender planning tools that are specific to the circumstances of the area. Creating appropriate methodologies to include gender at every stage of a project cycle, from design to implementation to evaluation, is another significant contribution that development organizations can make to address the marginalization of women in the development process.

Non-governmental organizations are another significant channel for effort to address discrimination against women. These NGOs range from well-organized and well-funded establishments to informal voluntary groups. NGOs provide a wide range of services to women and girls including various forms of education including literacy class and the training on income-generating skills. Other NGOs have offered health services and information. Still others have encouraged formal political participation among women, and have educated women about their legal rights, especially regarding family law. Some of the more substantial NGOs have engaged in political lobbying for the rights of women, and have launched awareness campaigns. These forms of political activism are essential: they ensure political pluralism and can often reflect and communicate the perspectives of their female constituencies. Measures to strengthen NGOs and to allow a wider breadth of movement are essential to any sincere efforts to eradicate gender discrimination.

Government also has a critical role to play in resolving some of the structural obstacles that prevent women from reaching their full potential, and that as a result, undermine the process of development as a whole. For example, increasing literacy rates among women and raising the level of female education should be a central priority for government throughout the Middle East and North Africa. Widening the pool of skilled labor can only add momentum to region's expanding economies, and its populations can only benefit from the higher health standards linked with women's education. Additionally, the governments of the region have the responsibility to review legal codes that are gender-biased and to establish feasible venues of recourse for women who are discriminated against in their workplaces or in their homes.

Finally, bringing gender to the attention of the mass media and support for more sophisticated and in-depth coverage of gender-related issues is possibly the most effective strategy for redressing gender inequity. Involving the mass media inserts gender into the public debate, and can a create an environment that fosters and even compels policy reform. Creating possibilities for women to become equal partners in development can only make the process more sustainable and more rich. Our shifting growing economies demand a shifting broadening awareness of gender.

Media and communication

Background

With the advent of the age of information, the media can play an important role in raising public awareness and in fostering increased participation in public affairs. Politicians have, for a long time, recognized the importance of the mass media in mobilizing public support. Businesses have also appreciated the value of the mass media, investing heavily in the promotion of their image and products. Meanwhile, development planners and practitioners have invested very little in bringing to the public attention the issues underlying sustainable development. Popularizing development issues to the point where they can be effectively communicated through the mass media, and stimulating the interest of the media in sustainable development issues becomes, therefore, a major challenge facing civil society, government, and development organizations. If the strength of the media were used to rally support for the civic issues underlying sustainable development, it would mean a lot more public involvement and contribution.

Until now, development professionals have generally perceived the media as a tool for the dissemination of their agendas and selected information at best, and as an intruder unable to understand the complexities of community development challenges at worse. Instead of relying on carefully choreographed media campaigns, a move by development organizations to involve the media on equal footing is by far the most effective way of injecting development issues into public discussions, debate that can ultimately transform development priorities to more accurately reflect local needs and concerns. Such a move would undercut the imposition of unilateral or overly-directive agendas. It would also allow local communities to feel a sense of ownership of development projects, the best guarantee for project sustainability.

Approach

Mobilizing the mass media to draw the topic of development into the public domain is predicated on the ability of development planners to make sustainable development issues interesting to both the mass media and the general public. We see the need for a comprehensive approach that is made up of the following three parts: making development organizations more media savvy, building capacity of the media, and creating media networks in the region and beyond.

Development organizations would benefit if they began to perceive a strong connection with the media as an important priority, and if they invested in ameliorating their public relations and media relations skills. Working closely with media specialists would familiarize them with techniques for constructive interaction of the mass media. The most important of these skills is learning how to present development issues in a way that is both intriguing and accessible to the general public. This means trading in development jargon for language understandable to the lay person, and the use of a communication style that flows from the assumption that if sustainable development issues and approaches are presented in a clear and sensible manner, the wider public is perfectly capable of grasping and interrogating them.

In addition to overtures made by development organizations, a crucial measure for fostering more incisive coverage of sustainable development challenges is building the capacity of the media. As development issues become more complex, there is a need for a more sophisticated media that can discern and convey the impact of development on local and global levels. This kind of in-depth and subtle coverage requires that the media take full advantage of the new communication tools at their disposal. Equipping the media with the skills and instruments to access a large pool of information from around world with 21st century speed will enable them to offer the public true-to-life journalistic portraits of sustainable development, with all its intricacies and ambiguities. Providing media personnel with computer hardware, internet access, and the corresponding computer literacy skills as well as more traditional libraries is a strategic investment to improve their coverage and engage them more thoroughly in development issues.

Finally, to ensure that media personnel become the most effective communicators possible about the development process, it is essential to create networks of journalists at the national and regional levels. Such networks can be excellent forums for the exchange of information and experience and can only enrich media coverage. We have taken a number of steps in the region to establish such networks: we have supported the creation of media NGOs in Egypt, Yemen, and Jordan, and we plan to support the creation of similar NGOs in other countries throughout the region. The cooperative contacts and productive discussions that these networks spark result in coverage that is self-adjusting and self-improving. Members of these networks are forging new links and developing a new culture of critically-minded journalists that are more inquisitive and persevering in their search for answers, and more committed to disseminating information in order to bring sustainable development issues into the forefront of public consciousness.

Conclusion

As development agencies and governments are being challenged to tighten their fiscal belts, traditional approaches to development assistance deserve a careful review. Development organizations cannot feasibly continue to support their current levels of engagement in financially intensive training and technical assistance initiatives. Not only are they unaffordable in the long run, but their impact is also becoming increasing questionable. The time for traditional development assistance has passed: much has been spent and not enough has been accomplished.

Development organizations must adopt a new role that corresponds more closely to the needs of our times. They should invest more in catalyzing development processes and less in the design and implementation of technical assistance programs. They are likely to have the greatest impact when they act as trend-setters for private investment, providing seed capital around which private investment is mobilized. This strategy is the best guarantee of a continued source of financing for sustainable development in the region because it taps into the entrepreneurship of every individual, rich and poor, and into the wealth of the world, the bulk of which is concentrated in private hands.

Private investment, both inside and outside developing communities, is central to the expansion of markets and to the support of new entrepreneurs better equipped to service those markets. It will carve out new channels for the transfer of capital between countries of the North and South, so that funds for development will not have to be sifted through governments but can be injected directly into the local economy. Another benefit of this approach is that the scope and scale of development processes will not be limited by the capabilities or vision of development institutions. Instead, private investors will favor initiatives that show the requisite qualities for them to take root and spread. The role of government and development organizations will then be concentrated on removing barriers that interfere with trade and investment and that limit private efforts and community participation, as well as on reforming fiscal policies to encourage investment in high impact development initiatives and on supporting financial intermediation and technical cooperation programs that target business expansion in the small and micro enterprise sector and the environmental services industry.

The three main initiatives presented in this paper - new franchises for sustainable development, informal sector participation in environmental services, and the conservation of our natural assets and the revitalization of our historic city centers - are examples of development strategies that build on the vitality and ingenuity of private initiatives. They all display the promise for growth and expansion that attracts private investors. Most importantly, however, they will all build the economies of the region from the ground up, laying down a broad and diversified economic base and equipping the poor with the economic tools to break out of the cycle of poverty.

This approach to development in the region revives something old and offers something new. It calls on those of us in the Middle East and North Africa to peer back into our history and read the lessons encoded in our historic city centers. We are an urban civilization and it is in the heart of our cities that we find proof of the diversity of cultures, skills, backgrounds, and outlooks that intermeshed so gracefully. The cultural richness and the material prosperity that it brought to the region offers solid proof as to the value reorienting ourselves outward and in using our cities as platforms for promoting entrepreneurship, opening new markets and improving the efficiency development initiatives.

Reviving the hospitality that is so characteristic of the area will attract skills, ingenuity, and resources from the world over. Conserving our social graces and sense of humor will make our culture more accessible and our friendships more enduring. Returning to our tradition of looking outwards for opportunity and recognizing that our culture and civilization have never ceased to be interactive is nothing less than absolutely essential as we envision our future and plan out entry into the next millennium.

Note

* This paper was presented by Christian Arandel on behalf of Environmental Quality International (EQI).

11 Sustainable cities: a contradiction in terms?

Herbert Girardet

Introduction

City growth is changing the condition of humanity and the face of the earth. In one century, global urban populations will have expanded from 15 to 50 per cent, and this figure will increase further in the coming decades. By 2000, half of humanity will live and work in cities, while the other half will depend increasingly on cities and towns for their economic survival. In the UK, a pioneer of large scale urban development, over 80 per cent of people live in cities.

The size of modern cities, too, is unprecedented in terms of numbers as well physical scale: in 1800, there was only one city with a million people, London. At that time, the largest 100 cities in the world had 20 million inhabitants, with each city usually extending to just a few thousand hectares. In 1990, the world's 100 largest cities accommodated 540 million people and 220 million people lived in the 20 largest cities, mega-cities of over 10 million people, some extending to hundreds of thousands of hectares. In addition, there were 35 cities of over five million and hundreds of over one million people.[1]

In the 19th and early 20th centuries, urban growth was occurring mainly in the North, as a result of the spread of industrialization and the associated rapid increase in the use of fossil fuels. Today, the world's largest and fastest growing cities are emerging in the South, because of unprecedented urban-industrial development, and frequently as a consequence of rural decline.

We are used to thinking about cities as places where great wealth is generated, and also where social disparities and tensions have to be addressed. The urban social agenda is certainly a critical one, and much effort has gone into addressing these problems. Cities as cultural centers have also received much attention, with ancient cities the world over enjoying an unprecedented tourist boom, and great urban centers such as London, New York or Paris accepted as the epitome of cultural development. However, an issue which has received much less attention is the huge resource use of modern cities and the implications of that for both local and global environments.

World-wide, increased resource consumption is closely associated with urban development. This tends to cause a growth in human living standards in monetary terms which can be witnessed today in developing countries where urban people, typically, have much higher levels of consumption than rural dwellers, with massively increased through-put of fossil fuels, metals, timber, meat and manufactured products. However, increased resource throughput in urban centers is also a threat to the health of city people as they are exposed to high concentrations of disease vectors. Many third world cities do not have the infrastructures to cope with the accumulation of wastes. Diseases such as cholera, typhoid and TB, well known in London 150 years ago, are occurring in many developing cities, with epidemics threatening particularly the poorest districts.

In global environmental terms, too, increased resource use is becoming a pertinent issue. Asia is currently undergoing the most astonishing urban-industrial development. China alone, with 10 per cent economic growth per year, is planning to double the number of its cities, from just over 600 to over 1,200 by 2010. Some 300 million people are expected to be moving to cities, converting from peasant farming and craft-based living to urban-industrial lifestyles. The increased purchasing power is already leading to increased demand for consumer goods and a more meat based diet, with massive implications for the future environmental impact of the world's most populous country.[2]

Cities as superorganisms

Urban systems with millions of inhabitants are unique to the current age and they are the most complex products of collective human creativity. They are both organism and mechanism in that they utilize biological reproduction as well as mechanical production processes.

Large cities are evolving to have characteristics all of their own, and with the fine specialization and extraordinary diversity of skills, firms will tend to congregate where there is a large market - but the market is large precisely where firms' production is concentrated.[3] The vast array of productive enterprise, capital and labor markets, service industries and artistic endeavor could be described as a symbiotic cultural system. However, unlike natural systems they are highly dependent on external supplies: for their sustenance, large modern cities have become dependent on global transport and communication systems. This is not civilization in the old-fashioned sense, but mobilization, dependent on long-distance transport routes.

Demand for energy defines modern cities more than any other single factor. Most rail, road and airplane traffic occurs between cities. All their internal activities - local transport, electricity supply, home living, services provision and manufacturing - depends on the routine use of fossil fuels. As far as I am aware, there has never been a city of more than one million people not running on fossil fuels. Without their routine use, the growth of mega-cities of 10 million people and more would not have occurred. But there is a price to pay. Waste gases, such as nitrogen dioxide and sulfur dioxide, discharged by chimneys and exhaust pipes, affect the health of city people themselves and, beyond urban boundaries, forests and farmland downwind. A large proportion of the increase of carbon dioxide in the atmosphere is attributable to combustion in the world's cities. Concern about climate change, resulting mainly from fossil fuel burning, is now shared by virtually all the world's climatologists.

Concentration of intense economic processes and high levels of consumption in cities both increase their resource demands. Apart from a monopoly on fossil fuels and metals, humanity now uses nearly half the world's total photosynthetic capacity as well. Cities are the home of the "amplified man", an unprecedented amalgam of biology and technology, transcending his biological ancestors. Beyond their limits, cities profoundly affect traditional rural economies and their cultural adaptation to biological diversity. As better roads are built, and access to urban products is assured, rural people increasingly abandon their own indigenous cultures which are often defined by sustainable adaptation to their local environment. They tend to acquire urban standards of living and the mind set to go with these.

Urban agglomerations are becoming the dominant feature of the human presence on earth, with supplies brought in from an increasingly global hinterland. All in all, urbanization has profoundly changed humanity's relationship to its host planet, with unprecedented impacts on forests, farmland and aquatic eco-systems. The human species is changing the very way in which the "the web of life"[4] on earth itself functions, from a geographically distributed interaction of a myriad of species, into a punctuated system dominated by the resource use patterns of just the one species: cities take up only two per cent of the world's land surface, yet they use over 75 per cent of the world's resources.

With Asia, Latin America and parts of Africa now joining Europe, North America and Australia in the urban experiment, it is crucial to assess whether large scale urbanization and sustainable development can be reconciled. Whilst urbanization is turning the living earth from a self-regulating interactive system into one dominated by humanity, we have yet to learn the skill of creating a new, sustainable equilibrium.

London's footprint

A few years ago I produced a TV documentary on deforestation in the Amazon basin and the resulting loss of bio-diversity. Filming at the port of Belem, in Brazil, I saw a huge stack of mahogany timber with "London" stamped on it being loaded into a freighter. I started to take an interest in the connection between urban consumption patterns and human impact on the biosphere. It occurred to me that logging of virgin forests, or their conversion into cattle ranches and into fields of soya beans for cattle fodder (in Brazil's Mato Grosso region) or of manioc for pig feed (in the former rainforest regions of Thailand), was perhaps not the most rational way of assuring resource supply to urban "agglomeration economies".[5]

Recently, the Canadian economist William Rees started a debate about the ecological footprint of cities, which he defines as the land required to supply them with food and timber products, and to absorb their CO_2 output via areas of growing vegetation.[6] I have examined the footprint of London, which also happens to be the city that started it all: the "mother of mega-cities". Today London's total footprint, following Rees' definition, extends to around 125 times its surface area of 159,000 hectares, or nearly 20 million hectares (see Appendix 1). With 12 per cent of Britain's population, London requires the equivalent of Britain's entire productive land.[7] In reality, this land, of course, stretches to far-flung places such as the wheat prairies of Kansas, the tea gardens of Assam, the forests of Scandinavia and Amazonia, and the copper mines of Zambia.

But large modern cities are not just defined by their resource use. They are also centers for financial services. When discussing urban sustainability we have to try to assess the financial impact of cities on the rest of the world.

A friend of mine recently told me about a startling experience:

196

A couple of years ago, I attended a meeting typical of those which take place every day in the city of London. A group of Indonesian businessmen organized a lunch to raise £ 300 million to finance the clearing of a rainforest and the construction of a pulp paper plant. What struck me was how financial rationalism often overcomes common sense; that profit itself is a good thing whatever the activity, whenever the occasion. What happened to the Indonesian rainforest was dependent upon financial decisions made over lunch that day. The financial benefits would come to institutions in London, Paris or New York. Very little, if any of the financial benefits would go to the local people. Therefore, when thinking about the environmental impact of London, we have to think about the decisions of fund managers which impact on the other side of the world. In essence, the rainforest may be geographically located in the Far East, but financially it might as well be located in London's Square Mile.[8]

A crucial question for a world city such as London is how it can reconcile its special status as a global trading center with the new requirement for sustainable development. London's own development was closely associated with gaining access to the world's resources. How can this be reconciled with creating a sustainable relationship with the global environment and also with the aspirations of people at the local level?

London was a pioneer in large-scale urban development. Today London's businesses, its few remaining manufacturing companies as well as its trading corporations and its financial institutions, certainly have the desire to continue. They wish to be sustainable in their own right. The question now is how the momentum for sustainable development can encompass the aspirations of business, assuring that people can lead lives of continuity and certainty, whilst together achieving compatibility of the urban metabolism with the living systems of the biosphere.

The metabolism of cities

Like other organism, cities have a definable metabolism. The metabolism of most "modern" cities is essentially linear, with resources flowing through the urban system without much concern about their origin, and about the destination of wastes; inputs and outputs are considered as largely unrelated. Raw materials are extracted, combined and processed into consumer goods that end up as rubbish which cannot be beneficially reabsorbed into living nature. Fossil fuels are extracted from rock strata, refined and burned; their fumes are discharged into the atmosphere.

In distant forests, trees are felled for their timber or pulp, but all too often forests are not replenished. Similar processes apply to food: nutrients that are taken from the land as food is harvested, and not returned. Urban sewage systems usually have the function of collecting human wastes and separating them from people. Sewage is discharged into rivers and coastal waters down stream from population centers, and is usually not returned to farmland. Today coastal waters are enriched both with human sewage and toxic effluents, as well as the run-off of mineral fertilizer applied to farmland feeding cities. This open loop is not sustainable.

The linear metabolic system of most cities is profoundly different from nature's circular metabolism where every output by an organism is also an input which renews and sustains the whole living environment. Planners designing urban systems should start by studying the ecology of natural systems. On a predominantly urban planet, cities will need to adopt circular metabolic systems to assure the long-term viability of the rural environments on which they depend. Outputs will also need to be inputs into the production system, with routine recycling of paper, metals, plastic and glass, and the conversion of organic materials into compost, returning plant nutrients to keep farmland productive.

The local effects of urban use of resources of cities also needs to be better understood. Cities accumulate large amounts of materials within them. Vienna, with 1,6 million inhabitants, every day increases its actual weight by some 25,000 tons.[9] Much of this is relatively inert materials, such as concrete and tarmac. Other materials, such as heavy metals, have discernible environmental effects: they gradually leach from the roofs of buildings and from water pipes into the local environment. Nitrates, phosphates or chlorinated hydrocarbons accumulate in the urban environment and build up in water courses and soils, with as yet uncertain consequences for future inhabitants.

The critical question today, as humanity moves to full urbanization, is whether living standards in our cities can be maintained whilst curbing their local and global environmental impacts. To answer this question it helps to draw up balance sheets comparing urban resource flows (see Appendix 2). It is becoming apparent that similar-sized cities supply their needs with a greatly varying throughput of resources. Most large cities have been studied in considerable detail and in many cases it will not be very difficult to compare their use of resources. The critical point is that cities and their people could massively reduce their throughput of resources, maintaining a good standard of living whilst creating much needed local jobs in the process.

Are solutions possible?

It seems unlikely that the planet can accommodate an urbanized humanity which routinely draws its resources from a distant hinterland. Can cities therefore transform themselves into sustainable, self-regulating systems - not only in their internal functioning, but also in their relationships with the outside world? An answer to this question may be critical to the future well-being of the biosphere, as well as of humanity. Maintaining stable linkages with the world around them is a completely new task for city politicians, administrators, business people and people at large. Yet there is little doubt that the world's major environmental problems will only be solved through new ways of conceptualizing and running our cities, and the way we lead our urban lives.

Today we have the historic opportunity to implement technical and organizational measures for sustainable urban development, arising from agreements signed by the international community at UN conferences in the 1990s. Agenda 21 and its prescriptions for solving global environmental problems at the local level are well known. Building on Agenda 21, the Habitat Agenda, signed by 180 nations at the 1996 Habitat II conference in Istanbul, will also strongly influence the way we run cities. It states:

> Human settlements shall be planned, developed and improved in a manner that takes full account of sustainable development principles and all their components, as set out in Agenda 21. ... We need to respect the carrying capacity of ecosystems and preservation of opportunities for future generations. Production, consumption and transport should be managed in ways that protect and conserve the stock of resources while drawing upon them. Science and technology have a crucial role in shaping sustainable human settlements and sustaining the ecosystems they depend upon.[10]

What, then, is a sustainable city? Here is a provisional definition: a "sustainable city" is a city that works so that all its citizens are able to meet their own needs without endangering the well-being of the natural world or the living conditions of other people, now or in the future. This definition concentrates the mind on fundamentals. In the first instance the emphasis is on people and their needs for long term survival. Human needs include good quality air and water, healthy food and good housing; they also encompass quality education, a vibrant culture, good health care, satisfying employment or occupation, and the sharing of wealth; as well as factors such as safety in public places, supportive relationships, equal opportunities and freedom of expression; and meeting the special needs of the young, the old or the disabled. In a sustainable city, we have to ask: are all its citizens able to meet these needs?

Conditions for sustainable development

Given that the physiology of modern cities is currently characterized by their routine use of fossil fuels to power production, commerce, transport, water supplies as well as domestic comfort, a major issue for urban sustainability is whether renewable energy technologies may be able reduce this dependence. London, for instance, with 7 million people, uses 20 million tons of oil equivalent per year, or two supertankers a week, and discharges some 60 million tons of carbon dioxide. Its per capita energy consumption is amongst the highest in Europe, yet the know-how exists to bring down these figure by between 30 and 50 per cent without affecting living standards, whilst creating tens of thousands of jobs in the coming decades.[11]

To make them more sustainable, cities today require a whole range of new resource efficient technologies, such as combined heat-and-power systems, heat pumps, fuel cells and photovoltaic modules. In the near future enormous reductions in fossil fuel use can be achieved by the use of photovoltaics. According to calculations by BP, London could supply most of its current summer electricity consumption from photovoltaic modules on the roofs and walls of its buildings.[12] This technology is still expensive, but large scale production will massively reduce unit costs.

Looking back, the physiology of traditional towns and cities was defined by transport and production systems based on muscle power. Dense concentration of people was the norm. Many cities in history adopted symbiotic relationships with their hinterland to ascertain their continuity. This applies to medieval cities with their concentric rings of market gardens, forests, orchards, farm and grazing land. Chinese cities have long practiced the return of nightsoil onto local farmland as a way of assuring sustained yields of foodstuffs.[13] Today most Chinese cities administer their own, adjacent areas of farmland and, until the recently, were largely self-sufficient in food.[14]

A major effect of the routine use of fossil-fuel based technologies was for cities to replace this density with urban sprawl. Motor transport has

200

caused many cities to stop relying on resources from their local regions and to become dependent on an increasingly global hinterland. However, some modern cities have made circularity and resource efficiency a top priority. Cities right across Europe are installing waste recycling and composting equipment. Austrian, Swiss, Danish and French cities have taken the lead. In German towns and cities, at this point in time, dozens of composting plants are under construction. Throughout the developing world, too, cities have made it their business to encourage recycling and composting of wastes.[15]

Some writers have argued that cities can actually be beneficial for the global environment, given the reality of a vast human population.[16] They suggest that the very density of human life in cities makes for energy efficiency in home heating as well as in transport. Systems for waste recycling are more easily organized in densely inhabited areas. And urban agriculture, too, if well developed, could make a significant contribution to feeding cities and providing people with livelihoods.

Urban food growing is certainly common in the late 20th century and not just in poorer countries, and a new book published by UNDP proves the point, as the following data indicates. The 1980 US census found that urban metropolitan areas produced 30 per cent of the dollar value of US agricultural production. By 1990, this figure had increased to 40 per cent. Singapore is self-reliant in meat and produces 25 per cent of its vegetable needs. Bamako, Mali, is self sufficient in vegetables and produces half or more of the chickens it consumes. Dar-es-Salaam, one of the world's fastest growing large cities, now has 67 per cent of families engaged in farming, compared with 18 per cent in 1967. In Moscow, 65 per cent of families are involved in food production, compared with 20 per cent in 1970. There are 80,000 community gardeners on municipal land in Berlin, with a waiting list of 16,000.[17]

Policies for sustainability in the UK

Today we have a great opportunity to develop a whole new range of environmentally friendly technologies for use in our cities. Efficient energy systems are now available for urban buildings, including combined heat-and-power generators, with fuel cells and photovoltaic modules waiting in the wings. New materials and concepts of architectural design allow us to greatly improve the energy performance and to reduce the environmental impact of materials use in buildings. Waste recycling technologies for small and large, rich and poor cities, can facilitate greater efficiency in the urban use of resources. Transport technologies, too, are due for a major overhaul. Fuel efficient low emission vehicles are at a very advanced stage of development. In US cities, rapid urban transit systems are starting reappear even where people had come to depend almost exclusively on private transport.

With over 80 per cent of the population in the UK living in cities, and with cities using most of the world's resources, it is critically important to develop new policies for sustainability - social, economic and environmental. Britain has the historic opportunity of developing practical policies for urban redevelopment, benefiting both people and the environment. This means, above all else, self-financing investment in end use efficiency, reducing resource use whilst simultaneously generating urban jobs and business opportunities. With Britain signed up to Agenda 21 and the Habitat Agenda, we have the opportunity to refocus investment from resource extraction to resource conservation and recycling, with a great many employment and business opportunities. Whilst a policy based on high resource productivity would reduce employment in mining, much of it abroad, it would enhance job creation in end use efficiency - in the building trade, in environmental technology industries and in the electronics sector - in places where they are most needed: in our cities.

Policies proposed here aim to create synergies between various business sectors: the waste outputs of cities can be a basis for new business ventures. Energy efficiency, so far tackled half-heartedly, should be give top priority. Government can do a great deal to facilitate sustainable urban development, using European and national legislation, planning regulation and budgetary signals to initiate change.

Waste

The recent landfill tax is increasing recycling in the UK, helping to achieve the government target of 25 per cent household waste recycling by 2000. This taxation should be extended, with the purpose of achieving a recycling rate of at least 50 per cent which is already the norm in other countries. In some British cities, such as Bath and Leicester, where recycling has advanced a great deal, the benefits for people and the local environment are already clearly in evidence.

In London - where currently only seven per cent of household waste is recycled - a new initiative by LPAC and London Pride Partnership is expected to bring recycling up to unprecedented levels: by 2000, every London home will have a recycling box with separate compartments instead of conventional dustbins. Progressively more and more municipal waste will be recycled, establishing new reprocessing industries and creating 1,500 new jobs.[18] This figure will go up further early in ten new century. Already composting organic wastes is advancing well, with "timber stations" composting shredded branches of pruned trees and leaf litter being established in various locations.

Energy

National planning regulations have already greatly improved energy efficiency of homes, but much more can be done, dramatically improving the energy performance of buildings, creating more local jobs in the process and reducing environmental impact of energy use. Regulating the energy supply industry to further improve generating efficiency and reduce discharge of waste gases, could significantly increase the use of modern, clean CHP systems in our cities.

The UK is just seeing the first schemes where greenhouse cultivation is being combined with CHP, utilizing their hot water and waste CO_2 to enhance crop growth for year-round cultivation.[19] Policies encouraging CHP could thus also be used for enhancing urban agriculture, bringing producers closer to their markets instead of flying and trucking in vegetables long distance. Once again, beneficial effects on jobs would result.

Another policy area is photovoltaics. Governments policy should vigorously encourage the installation of photovoltaic modules on buildings, enhancing the UK's capacity to produce PV systems and creating much needed local jobs in the process. Experimental buildings such as the photovoltaics center at Newcastle University, a 1960s' building recently clad with photovoltaic panels, are very promising.[20] Every city in the UK should have such buildings to test the potential of PV and to develop local know-how.

Sewage

A major urban output is sewage, containing valuable nutrients such as nitrates, potash and phosphates. Returning these to the land is an essential aspect of sustainable development. In Bristol, Wessex Water now dries and granulates all of the city's sewage. The annual sewage output of 600,000 people is turned into 10,000 tons of fertilizer granules. Most of it is currently used to re-green the slag heaps around Merthyr Tydfil across the Severn in South Wales. In contrast, Thames Water in London is currently constructing incinerators for burning the sewage sludge produced by four million Londoners. This is a decision of historic short-sightedness given that phosphates - only available from North

Africa and Russia - are likely to be in short supply within decades. Crops for feeding cities cannot be grown without phosphates.

Cities all over the UK should be encouraged to build sewage recycling works using the latest technologies such as the one utilized by Wessex Water. This will be of critical importance for sustainable urban development and will, once again, create jobs in cities as well as benefiting the environment.

Smart cities

Cities are centers of communication and new electronic systems have dramatically enhanced that role. Information technologies have given cities a global reach as never before, and particularly in further extending the financial power of urban institutions. The daily money-go-round from Tokyo to London and on to New York and Los Angeles is the most striking example of this, as the new economy is organized around global networks of capital, management, and information, whose access to technological know-how is at the roots of productivity and competitiveness.[21] But will this power ever be exercised with a sense of responsibility appropriate to an urban age? If this is the global network society, who controls its ever growing power?

The global economic and environmental reach of cities today needs to be matched with communication systems that monitor new impacts, an early warning system that enables city people to ring alarm bells as soon as new, unacceptable developments occur, whether it is the transfer of toxic waste or the transfer of environmentally undesirable technologies from one city or territory to another.

Much more needs to be done to ensure processes by which cities monitor and ameliorate their impact on the biosphere. I would like to postulate that modern cities could develop cultural feedback systems, responding to the challenge of achieving sustainability by limiting urban resource consumption and waste output through technological and organizational measures. Today new communication technologies should also be utilized to improve the functioning of cities in many different ways, and communications within them. Urban Intranets, now in place in a growing number of cities, should enhance the communication flow between various sectors of urban society. Electronic sampling of opinions should be used to enhance urban decision making.

In that context it is of critical importance to recognize the great inherent creativity of city people. In the end only people can implement measures for sustainable urban development - technical fixes are not enough. But people need a good knowledge base. For this purpose, the most important thing is the collection and dissemination of best practices to assure that people in cities world-wide actually know about existing projects. Much better use has to be made of new information technologies. That would be an indication that cities were becoming smart in the best sense of the word.

The legacy of Habitat II

The 1996 Habitat II Conference in Istanbul made a great deal of the fact that cities, more often than not, are considered places where problems are concentrated; yet, in reality, given half a chance, people wherever they are, seek to improve their situation wherever possible. The Best Practices and Local Leadership Program was one of the flagship initiatives of Habitat II. In the course of two years, it collected some 700 examples from around the world under the following categories: poverty reduction and job creation; access to shelter, land and finance; production/consumption cycles; gender and social diversity; enterprise and economic development; waste re-cycling and re-use; transport and communication; combating social exclusion; crime prevention and social justice; and governance. It also comprises the following categories: infrastructure, water and energy supply; innovative use of technology; environmental protection and rehabilitation; and last, but not least, policy and planning. This information is now available via e-mail[22] and through direct contacts with urban groups all over the world. Exchange programs for disseminating this information are now reaching some of the poorest urban communities.

In some cities, too, business is recognizing the need to make urban sustainability a central concern and to support local communities, but much more needs to be done. According to Wally N'Dow, Director-General of Habitat II, there are five lessons that emerged out of the preparations for the Conference:

1 The power of the good examples: there are fascinating initiatives throughout the world's cities. Habitat and its partners have helped groups from around the world to prepare reports and to make films about their own activities. It is also undertaking the dissemination of best practices. This process will deepen our understanding of urban challenges and opportunities so that realistic steps can be taken at local, national and international levels to develop new partnerships for solving problems and enriching the life of cities.

2 Complexity of issues: the contributions Habitat received also illustrated just how complex modern cities are. In this context, obstacles to successful implementation must be analyzed and effective processes for implementing projects identified. In situations of rapid urban growth it is particularly important for the development of urban infrastructure problems to be overcome.

3 Local level action has large scale repercussions: implementation must be tailored very closely to local situations. We then have to ask: how applicable are best practices outside their own regions? For urban best practice to be transferable from one city to another, implementation must be closely tailored to local situations. It is particularly important to establish under what circumstances and with what types of partners successful projects have materialized.

4 Exchanges take place between peer groups in different cities: the sharing of best practice between cities is an essential tool for sustainable urban development. Once outside interest in a project has been established, site visits are of critical importance. By learning from example, local transformation can lead to global change.

5 Changing the way urban institutions work: the power of allowing people direct access to best practice examples through a dynamic process of decentralized co-operation has become very apparent. The material collected under the Habitat "best practice initiatives" is a gold mine for the world's cities and its dissemination will be of paramount importance for all the potential partners concerned. [23]

By the time the next century passes its first quarter, more than a billion and a half people in the world's cities will face life and health threatening environments unless we create a revolution in urban problem solving. We need a new approach, a creative and constructive effort that can only come if we forge a global partnership between national governments and local communities, between the public and private sectors. Whether it is the environment or human rights, population or poverty or the status of women, we must deal with these issues in our cities. That is why they have become a priority challenge for the international community, and why it is essential that they are at the center of a growing global effort to make our cities and all communities productive, safe, healthy, more equitable and sustainable.

To make a success of cities we need to extend popular participation in decision-making to restore confidence in local democracy. Consultation is not enough. To strengthen local democratic processes, methods such as such as neighborhood forums, action planning and consensus-building should be widely used, because in appropriate circumstances these lead to better decisions and easier implementation. With the help of modern communications technologies, wider citizens' involvement can be incorporated into strategic decision-making.

Cultural development

With the whole world now copying western development patterns we need to formulate new cultural priorities. Cultural development is a critical aspect of sustainable urban development, giving cities the chance to realize their full potential as centers of creativity, education and communication. Cities are nothing if not centers of knowledge and today this also means knowledge of the world and our impact on it. Reducing urban impacts is as much an issue of education and of information dissemination as of the better uses of technology.

Ultimately, that cannot be done without changing the value systems underpinning our cities. Adopting circular resource flows will help cities reduce their footprint, and thus their impact on the biosphere is a cultural issue. Initiatives to that effect are now in evidence all over the world. In many cities there is growing awareness that the urban super-organism can become a sustainable, self-regulating system through appropriate cultural processes. In the end, it is only a profound change of attitudes, a spiritual and ethical change, that can bring the deeper transformations that can make cities truly sustainable.

We need to revive the vision of the city as a place of culture and creativity, of conviviality and above all else of sedentary living. As I have suggested, currently cities are not centers of civilization but mobilization of people and goods. A calmer, serener vision of cities is needed to help them fulfill their true potential as places not just of the body but of the spirit. Great cities of the past were above all else places of beauty, with their great public spaces, their magnificent bridges and the rising spires of their religious buildings.

Cities are what their people are. The greatest energy of cities should flow inwards, to create masterpieces of human creativity, not outwards, to bring in ever more products from ever more distant places. The future of cities crucially depends on the utilization of the rich knowledge of their people, and that includes environmental knowledge. Cities cannot claim to be knowledge-led without activating the know-how to beautify their own internal environment for all to enjoy and to reduce their impact on environments world-wide. Central and local governments are increasingly aware that efforts to improve the living environment must focus on cities. Eco-friendly urban development could well become the greatest challenge of the 21st century, not only for human self-interest, but also for the sake of a sustainable relationship between cities and the biosphere on which humanity ultimately depends.

Cities for a new millennium will be energy and resource efficient, people friendly, and culturally rich, with active democracies assuring the best uses of human energies. In Northern mega-cities, such as London and New York, prudent inward investment will contribute significantly to achieving higher levels of employment. In cities in the South, significant investment in infrastructure will make a vast difference to health and living conditions. But none of this will happen unless we create a new balance between the material and the spiritual, and to that effect much good work needs to be done in the years to come.

A new paradigm

Thought has created the unstable world in which we now live - manifested in mega-technology, mega-cities, global power structures and vast environmental impacts. Practical visions and working examples of innovative, alternative systems are now urgently needed. We urgently need new thinking on sustainability, peace and personal empowerment. By emphasizing human scale solutions we can contribute to a core transformation of contemporary urban culture. We now need to develop concepts for real sustainability that can help bring about the reconciliation between people and between cities and nature.

These need to involve the whole person (mind, spirit and body); empower the individual for the good of the community; act locally keeping in mind the interests of the global community; place long term stewardship above short term satisfaction; ensure justice and fairness in a context of responsibility and sharing; identify the appropriate scale of viable human activities. They also need to encourage diversity within the unity of a given community; draw from the past, learn from the present and face up to the future; ascertain precautionary principles to determine the effects of our actions; and assure that our use of resources does not diminish the living environment.

The implementation of such ideas in an urban context is long overdue.

Notes

1 See UNCHS (1996).
2 See Worldwatch Institute Washington (1997).

3 Data provided by LSE Greater London Group (1996).
4 See Capra (1996).
5 As in 3.
6 See Wackernagel and Rees (1996).
7 See Girardet (1996a).
8 Mark Campanale, personal communication.
9 Professor Paul Brunner, TU, Vienna, personal communication.
10 See UN (1996).
11 As in 7.
12 Rod Scott, BP Solar, personal communication.
13 See King (1911).
14 See Sit (1988).
15 According to the *Warmer Bulletin*, Summer 1995.
16 See Gilbert et al (1996).
17 See UNDP (1996).
18 *Evening Standard*, 30 December 1996.
19 According to the *Grower Magazine*, 21 March 1996.
20 See Hill (1996).
21 See Castells (1996).
22 The website is http://www.bestpratices.org
23 See Girardet (1996b).

References

Capra, F. (1996) *The Web of Life*, Harper Collins: London.
Castells, M. (1996) *The Network Society*, Blackwells: Oxford.
Gilbert, R. et al (1996) *Making Cities Work: the role of local authorities in the urban environment*, Earthscan: London.
Girardet, H. (1996a) *Getting London in Shape for 2000*, London First: London.
Girardet, H. (1996b) *The Gaia Atlas of Cities*, Gaia Books: London.
Hill, R. (1996) *Northumbria Solar Project*, University of Newcastle.
LSE Greater London Group (1996) *London's Size and Diversity*, January.
King, F.H. (1911) *Farmers of Forty Centuries*, Rodal Press: Emmaus.
Sit, V. (ed) (1988) *Chinese Cities: the growth of the metropolis since 1949*, Oxford University Press: Oxford.
United Nations (1996) *The Habitat Agenda*, UN: New York.
United Nations Center for Human Settlements-UNCHS (1996) *An Urbanizing World - Global Report on Human Settlements*, Oxford University Press: Oxford.

United Nations Development Program-UNDP (1996) *Urban Agriculture*, UN: New York.

Wackernagel, M. and Rees, W. (1996) *Our Ecological Footprint*, New Society.

Worldwatch Institute Washington (1997) *State of the World*, Earthscan: London.

Appendix 1
London's footprint

Population: 7,000,000 people

Surface area: 158,000 ha

Area required for food production: 1.2 ha per person: 8,400,000 ha

Forest area required by London for wood products: 768,000 ha

Land area that would be required for carbon sequestration = fuel production: 1.5 ha per person: 10,500,000 ha

Total London footprint: 19,700,000 ha = 125 times London's surface area

Britain's productive land: 21,000,000 ha

Britain's surface area: 24,400,000 ha

Source: compiled by Herbert Girardet, 1996

Appendix 2
The metabolism of Greater London, population 7,000,000

Inputs	*Tons per year*
Total tons of fuel, oil equivalent	20,000,000
Oxygen	40,000,000
Water	1,002,000,000
Food	2,400,000
Timber	1,200,000
Paper	2,200,000
Plastics	2,100,000
Glass	360,000
Cement	1,940,000
Bricks, blocks, sand and tarmac	6,000,000
Metals (total)	1,200,000

Wastes	
CO_2	60,000,000
SO_2	400,000
NO_X	280,000
Wet, digested sewage sludge	7,500,000
Industrial and demolition wastes	11,400,000
Household, civic and commercial wastes	3,900,000

Source: compiled by H. Girardet, 1995 and 1996